PRAISE FOR *CUSTOMER FOCUSED PROCESS INNOVATION*

"Every now and then a book comes along that forces us to rethink how we collectively organize our resources and manage our strategic initiatives. *Customer Focused Process Innovation* is exactly such a book. Refreshingly, David Hamme's approach is not a pie-in-the-sky set of big ideas, but rather an expertly woven and intuitively practical framework of concepts and tools that bridge the gap between visions and strategies, and their realization in the factories, stores, and other locations where corporately conceived value propositions are brought to life. It is simply a must-read for executives challenged with delivering,"

—Steve Jegier, Head of Strategy, Wealth, Brokerage, and Retirement, Wells Fargo

"David Hamme has created an important resource for executives and leaders who know that they must make fundamental changes in their organizations. The methodology of customer-focused strategies in this guide provides an invaluable blueprint to reinvent how your company innovates."

—Marshall Goldsmith, author or editor of 34 books including the global bestsellers *MOJO and What Got You Here Won't Get You There.*

"Dave Hamme brings energy, creativity, and over 20 years of converting strategy to action to his new book *Customer Focused Process Innovation.* His perspective on seeing the opportunities and then systematically designing your organization to deliver against them resonated with my experience in leading large innovation focused organizations from Shanghai to the midwest of the United States. I have known Dave for 20 years as a guy who can both think and deliver. This book guides us in how to do the same."

—Dave Ricks, Senior Vice President and President of Lilly Bio-Medicines

"Ninety-nine percent of the innovation books published don't address the part of the process where ninety-nine percent of the value is created: implementation. However, David addresses this complex issue head on. If you want to make innovation a reality, get this book."

—Stephen Shapiro, author, *Best Practices are Stupid*

"Hamme takes a microscope to the intricate machinery that drives whether a company succeeds or not—its processes and the people that perform them. Throughout, his emphasis on using a process focus and initiative management to drive long-term, real improvement to an organization is spot on ... it's the only way to really bring about the adaptation needed to survive. With an executive perspective but a practitioner's passion, his work can help mid- and senior-level leaders to take a new look at building resiliency into their operations."

—Richard Maltsbarger, Business Development Executive,
Lowe's Companies, Inc.

"When I say the words 'process' and 'innovation' I imagine two opposite ends of a spectrum: humdrum processes and exciting innovation! But if these two could synch up, how powerful would that be! Dave Hamme, a dear and trusted colleague for many years, has put these concepts together and now shows us how to instill excellence in our staff and organizations. Having a 'great idea' is easy—driving it to execution is the hard part. We all talk about it, and now Dave has it documented. Kudos, Dave!"

—Caroline M. Kolman, PE, Managing Director,
Prism Healthcare Partners LTD

"It is a rare treat to have practical yet very strategic advice laid out in such a digestible manner. *Customer Focused Process Innovation* brings a fresh perspective to some very well studied concepts in process improvement and innovation. David speaks from the perspective of a well-seasoned management consultant who doesn't just talk about these concepts, he implements them successfully. I had the pleasure of working with David on one of the most impactful efficiency programs of my career."

—Mike Lee, President, North Highland Worldwide Consulting

"Nonprofit organizations and for profit entities share many organizational characteristics. As nonprofits face unrelenting scrutiny, there is a growing focus on efficiency and productivity. *Customer Focused Process Innovation* is a playbook in creating a blueprint to unleash innovation, build partnerships and drive performance. Dave Hamme brings a wealth of information and savvy to a topic of great value to leaders of organizations of all sizes."

—Anne Pfeiffer, founder and Executive Director,
Pat's Place Child Advocacy Center

"Dave Hamme has taken the complexities of a business organization and developed a process-based framework to gain understanding and identify business opportunity. His process innovation builds from the bottom up in a practical, realistic manner that cannot only be aligned with the business strategy but also with the organization's ability to execute. Dave's book is a valuable read for any business leader trying to gain an edge in today's hyper-competitive business environment."

—Steve Buecking, Vice President and Executive Producer,
GA Communication Group

"Finally a business book that not only provides great thought provoking insight into how a business does or does not operate, but David Hamme also gives the step-by-step instructions to create a blueprint that everyone in the organization can follow to take the guesswork out of execution. *Customer Focused Process Innovation* is cutting edge common sense"

—Don Smith, Vice President of Marketing,
Family Dollar Stores

"I've been fortunate to rely on Dave in a professional setting as well as serving beside him in the non-profit world. Having observed, and benefitted from, his tremendous insight and innovation regarding business strategy and organizational behavior, I'm thrilled that he's going to share his approach and ideas. His book, *Customer Focused Process Innovation*, will be required reading for my entire management team so that we can continue to grow and evolve into a more effective business unit."

—Steve Newmark, President, Roush Fenway Racing

"Social media is pressuring businesses to react to customer sentiments and business opportunities in real time. Mobile and cloud based computing is enabling field based teams to react efficiently. All of this needs to be orchestrated through well thought out and proven processes and a customer-centric organization. This book couldn't have come at a better time."

—Chetan Saiya, CEO, Zoomifier Corporation

"David Hamme's book is spot on and insightful in regards to both the idea and the how to implement and gain alignment with 'Customer Focus,' 'Innovation' and 'Process'. Dave dives straight in by starting out on the process elements relevant to win. Through well-articulated principles he ties utilizing 'process' to be 'customer focused' extremely well. This core and essential theme is done with a connect-the-dots thinking that makes it easier to think about how to plow through the biggest challenges, maximize relevant innovation, get cost and quality right and have the speed to market essential to meeting customers' needs. The step-by-step approach he lays out transforms the complexity of process to a palatable idea you'll want to implement immediately."

—Gino Biondi, Vice President Sales and Marketing,
Zenith Products Corporation

CUSTOMER FOCUSED

FOCUSED

PROCESS

INNOVATION

CUSTOMER FOCUSED PROCESS INNOVATION

Linking Strategic Intent to Everyday Execution

David Hamme

New York   Chicago   San Francisco   Athens   London   Madrid
Mexico City   Milan   New Delhi   Singapore   Sydney   Toronto

1 2 3 4 5 6 7 8 9 0   DOC/DOC   1 2 0 9 8 7 6 5 4

ISBN   978-0-07-183470-4
MHID   0-07-183470-2

e-ISBN 978-0-07-183471-1
e-MHID   0-07-183471-0

Library of Congress Cataloging-in-Publication Data

Hamme, David.
   Customer focused process innovation : linking strategic intent to everyday execution / by David Hamme.
      pages cm
   ISBN 978-0-07-183470-4 (hardback) — ISBN 0-07-183470-2 (hardback)
   1. Reengineering (Management) 2. Organizational effectiveness.
   3. Organizational change. 4. Diffusion of innovations.
   5. Industrial management. I. Title.
   HD58.87.H35 2015
   658.4'063—dc23   2014015837

To River and Winter—for all the love and happiness you have brought to my life.

Contents

Preface

One of the great perks of a consulting career is the opportunity to work as a pseudo corporate tourist, examining the innards of dozens of companies and observing firsthand their operational practices and unique manner of conducting business. At each stop along the way, I watched and learned—discovering the attributes that characterized each enterprise and evaluating their influence on the enterprise's ability to achieve success. In every case, I discovered that there were always things the company did well . . . and there was usually ample opportunity for improvement. My experience hopping from client to client expanded my awareness of enterprise operations, cultures, processes, customer connections, resource management, and other business areas. This understanding vastly elevated my effectiveness with future clients. What later clients occasionally hailed as sage advice was simply parroting what I had learned previously.

Throughout my time on consulting engagements and in full-time roles, several commonalities surfaced that forced me to question some of the prevailing business structures and practices in existence at the time. First off, I was continuously surprised at how little leaders and top decision makers knew about the operational processes performed daily by their direct reports. A story passed to me during my early consulting years illustrates this condition.

From the 1960s up to its acquisition by Compaq in 1998, Digital Equipment Corporation (DEC) was a leading producer of computer systems, software, and peripherals. As the story goes, a meeting of senior leaders was convened after a massive reorganization failed to deliver the intended benefits. One of DEC's founders, Ken Olsen, presided over the meeting. After entertaining discussion for several hours, Mr. Olsen questioned his assembled team, "What you are telling me is that the senior managers in this company who are making commitments have no idea how this company operates?" At this point, there was dead silence until one of the senior executives who oversaw the reorganization bravely responded, "On an operational level, that is exactly what I'm telling you." Through my consulting engagements at over thirty companies, I know that this is far from an isolated incident.

While employed in Ernst & Young's process-improvement practice, I was staffed on roughly a dozen process-improvement engagements. The general approach was to map out a process, identify opportunities for improvement, build a future state, and implement the final solution. The knowledge I gained working on these projects was a goldmine. In nearly every case, the performance metrics proved that we were successful in improving the underlying process, but in the back of my head something nagged at me. We were missing something. The outcomes of our efforts did not appear to be substantially improving the company's financial results in the overwhelming number of instances. Shortly thereafter, I came to the realization that most process-improvement efforts are point solutions. That is, they improve a component of the enterprise but are essentially completed in isolation and without any discernible connection to the greater goals of the organization. Although they resulted in new efficiencies, the overall state of the enterprise remained basically unchanged. Even more unfortunate, these efforts consumed resources that might have been used more advantageously elsewhere in the enterprise.

As my career progressed and I ascended into executive positions, my doubts about the manner in which most large-scale projects were undertaken persisted. When asked to participate in driving strategic endeavors, I learned that creative ownership often fell to a handful of senior leaders who isolated themselves in a conference room to brainstorm solutions. Although on occasion these efforts resulted in solid opportunities, the majority had limited market validity, and very few moved the needle and improved the competitive position of the company. For the most part, my suspicions were confirmed that most improvement programs were birthed and executed in a relative vacuum without connection to a greater plan and without leveraging the full capabilities of the enterprise. In short, not only was the cannon aimed in the wrong direction, but a lot of powder was spilling onto the ground. What these companies were missing was a funnel for transformational efforts—a launching process that focused resources and energy where they could most benefit the enterprise.

This is the aim of this book. Over countless hours during which I served as a consultant and an executive, I struggled with the prevailing business practices and structures. They were unclear, misdirected, and largely preserved the status quo. After learning what didn't work, I set out on a quest to determine better ways of managing organizations and their responses to an ever-changing world. As my search progressed, it became evident to me that the optimal approach entailed the use of a familiar construct for defining work—*process*. And although previously considered primarily from the perspective of how to improve the efficiency of a component of work, processes are far more powerful than any of us think. Indeed, when designed, managed, and used correctly, processes are the foundational framework for innovation.

Acknowledgments

E ven in those instances when there is but one author's name listed on a book's cover, there is undoubtedly a sizeable population of individuals behind the scenes who left their imprint on the project through their encouragement, by providing invaluable guidance, or through their adoption of the concepts presented in the book. Were I to list all those individuals who aided me in organizing the pieces and putting them to paper, we would have a set of tomes rivaling Encyclopedia Britannica. This book was not created in a vacuum. Multitudes of clients and colleagues shared their ideas, collaborated on white boards, advised over a coffee, or worked elbow-to-elbow with me as these ideas took form. They deserve a large measure of any success that comes from the application of the concepts and tools presented herein. To all of them, I am eternally grateful.

In addition to these groups, there are three individuals deserving of special mention. First, my agent, John Willig of Literary Services Inc. John coached me throughout the book selling process and provided expert guidance. He's a true professional. I am extremely grateful for his support as I would still be a writer with just a lot of pages without him. Secondly, to my editor, Knox Huston. Knox has been a great ally throughout the process and this book is far better for having passed through his hands. Receiving a publication offer from McGraw-Hill has been a high point in my life. Thank you Knox. And finally, to my wife Allison. She has read every word I put to paper and has been a sounding board for all my ideas over the years. Her advice and counsel is baked into every chapter of this book. It simply would not exist without her. Thank you Allison!

1

The Challenge of Change

E very year, a growing number of corporate stalwarts—household
names anchored by esteemed brands—lose their footing and
stumble from what was previously considered an unassailable
market position. The past few years have been particularly unkind.
The list of failures added formerly admired brands such as Circuit
City, Woolworth's, Lehman Brothers, Borders Bookstores, Readers
Digest, Sharper Image, Washington Mutual, Bearing Point, Silicon
Graphics, Continental Airlines, and Wachovia, to name a few. Apart
from this list, a good number of companies remain open for business
only because either the U.S. taxpayer bailed them out or because
another company acquired them and retained the name.

At the helm of each of these failed companies was a leadership
team whose primary purpose was to deliver a financial return for
the company's stakeholders and to do so year after year. Then one
day these companies lost their edge and got beat. The market shifted
away from them, their customers sought something different, a new
competitor stole a chunk of their market share, or they simply failed
to keep abreast of the market. In a nutshell, they failed to innovate.
Outside of short term luck, innovation is the key to prosperity

in competitive markets. Every significant business venture traces its roots to an original spark of innovation. In most instances, the founder invented an experience or product with unique attributes that was superior to the available options. On occasion, the innovation was so ground breaking that it launched a brand-new market or fostered a new way of thinking. This is true of government institutions (think of our Founding Fathers), major companies (think of Bill Gates and Paul Allen of Microsoft), and other organizations as well (think of Nancy G. Bricker, who founded Susan G. Komen for the Cure). The founders of these enterprises were on a mission—to deliver products/services uniquely calibrated to their customers' needs. As a result of the ability of these organizations to create valued outcomes, their customers responded enthusiastically, and their names became recognizable to the masses. But there is a problem with reaching the top—once you arrive, it is increasingly hard to stay there.

Consider the widely quoted Dow Jones Industrial Average (DJIA), which tracks 30 of the largest and most widely owned public companies in the United States. Since its inception in 1896, the members of this honored group were chosen to mirror the ebb and flow of the greater business environment. Over the ensuing decades, many of these benchmark companies fell on hard times, were removed from the list, and replaced with a more suitable firm. Of the original 12 companies, General Electric is the sole survivor. While this alone is telling, consider that only five of the current members predate World War II, and a third of them were added since the late 1990s. Such turnover in the DJIA speaks to the challenge of continually performing in an ever-changing world.

Change manifests in many forms and continually evolves with the passage of time. To a corporate leader, change may be a promising market opportunity created by a newly passed legislative decree, or it may be an industry-wide upheaval precipitated by a scientific advancement. Its outcome may be minor tremors in a market, dissipating quickly and soon forgotten, or the change may be a monolithic disturbance that rewrites industry norms and sends even the most strategically impervious companies scrambling for cover. For example, consider the ebb and flow of several industries precipitated by major change events:

- *Banking industry.* As the financial markets plummeted during the Great Depression, the federal government's response launched the first of many upheavals in the financial industry. In 1933, the U.S. Congress passed a bill that became known as the Glass-Steagall Act. In order to limit the impacts of bank failures, this law limited the affiliations between commercial banks and securities firms. As decades passed and international banks expanded their operations, U.S. banks screamed that the act limited their global competitiveness by restricting the products and services they could offer their clients. Furthermore, financial institutions were finding clever ways to circumvent the intent of the law. In 1999, Congress, bowing to political pressure, passed the Gramm-Leach-Bliley Act, effectively repealing the Glass-Steagall restrictions. This legislation ushered in a flood of mergers and acquisitions at a pace that only slowed with the financial collapse of 2008.
- *Transportation industry.* In 1978, the U.S. Congress passed legislation to deregulate the airlines and allow fares to float based on market forces. Although arguably beneficial to passengers, the act initiated a series of price wars that resulted in a realignment of passenger routes. The unfortunate result was a series of bankruptcies across the major air carriers as they struggled to adapt. Southwest, a minor player back in 1978, prospered in the ensuing chaos by using well-chronicled innovations such as the elimination

of assigned seats and rapid gate turnarounds to limit the downtown of its planes. Since 2011, Southwest has flown more domestic passengers than any other airline on an annual basis.

- *Retail, entertainment, and publishing industries.* In the late 1990s, the Internet grew from a promising concept to a stable platform for publishing and conducting e-commerce. Retailers, newspapers, book publishers, music producers, and other industries struggled to adjust their business models to this new way of interacting with customers. Years later, they still continue to lose ground to former upstarts such as Amazon, eBay, Yahoo, Google, and Apple.

The point is that companies, industries, and even economies can be fundamentally altered in the blink of an eye. As Peter F. Drucker once said, "Managers may believe that industry structures are ordained by the Good Lord, but they can—and often do—change overnight."[1] No market, no product or service, and no company or enterprise is immune to the sudden barrage of disruptive forces that may well obliterate prevailing business structures and practices.

And to make things more interesting, the pace of change is accelerating. In an unprecedented manner, the Internet brings a deluge of information and product options directly into the customer's home. Having trouble making a purchasing decision? A host of experts and social acquaintances are available to offer their perspective on just about any purchase. And the selections available to consumers grow by the minute—especially given the increased presence of international firms. With comparatively cheap labor costs, significant untapped resources, ample capital, and the benefit of an expanding middle class in their own backyards, international companies are becoming fierce competitors not only in their home countries but also around the globe. Add to this the political, societal, technological, environmental, and other forces buffeting corporate America from every angle, and businesses are finding themselves engaged in an unending battle just to remain in the customer's consideration set. Businesses—in fact, enterprises of all forms—must adapt to the emerging realities if they wish to survive, let alone prosper. Change cannot be ignored. It continuously births obstacles to overcome and opportunities to exploit.

TO SURVIVE AND PROSPER REQUIRES INNOVATION

Enterprises that consistently succeed—those that seem to always react quickly to previously unforeseen opportunities and take the hill while the competition stands flatfooted—do so because they innovate. But what does *innovate* really mean? These days, *innovation* is a buzzword—strongly entrenched in the business vernacular but horribly overused and misused. It is increasingly difficult to put a finger on its meaning or even more so its applicability in the business world. In corporate America, most attempts to spur innovation center on conducting intense brainstorming sessions to elicit that hallowed epiphany—a moment of pure genius when a game-changing breakthrough is unearthed. Indeed, such an event is technically innovation—but expecting magical lucid moments to conveniently arrive is a fatalistic approach at best, the equivalent of betting an enterprise's future on the roll of dice.

From my experience, innovations are the result of a rigorous creation process in which an idea is tested and rebuilt over successive iterations until what remains is more right than the available alternatives. Howard Schultz built the concept that became Starbucks through repeated trial and error. His initial idea was to mimic the Italian coffee house experience he discovered while traveling abroad. Although the concept's initial performance was satisfactory, Schultz knew that there was room for improvement. So he refined the model to make it more appealing to American consumers. This included tinkering with the store's decor, redesigning the guests' passage through the store, and updating the assortment of products available. The result of his ongoing evolution of the Starbucks experience is a brand that has achieved worldwide prominence. Today Starbucks operates close to 19,500 stores in 62 countries and employs close to 200,000 people. Although Starbucks is frequently cited as an innovator, the overall concept was not born of a momentary blast of genius but rather developed through continual adjustments to a working model.

To the frustration of many, innovation is not easy. It is challenging on multiple levels. It requires two ingredients that are relatively scarce in established companies today. First, any innovation must be predicated on detailed and accurate information about the intended end state as well as what the enterprise is capable of producing. Which leads us to the second ingredient: innovation endeavors require a champion—an individual with the clout and awareness to shepherd meaningful change efforts. In most enterprises today, the sole position possessing the authority to get the full enterprise behind a major initiative is the chief executive officer (CEO). Unfortunately, these precursors for innovation spell its doom in most environments. Why is this case? Understanding the answer requires a brief review of the history of the modern organizational structure.

In *Reengineering the Corporation*, Michael Hammer and James Champy[2] trace the "work styles and organizational roots" of modern enterprises back to theories first presented by Adam Smith in *The Wealth of Nations*.[3] In particular, Hammer and Champy point to Smith's dissertation on the division of labor and how it perpetuated the use of specialists to manufacture products. During the industrial revolution, companies began segmenting workers based on their ability to complete a task. They learned that when individuals were assigned a single job, their efficiency increased as they became adept over continual iterations. By dividing the overall work effort into jobs performed by specialists, a product could be manufactured faster and cheaper. This finding led to the creation of divisions, departments, and other organizational delineations to manage these groups of specialized workers. As the manufacturing process gradually incorporated technology and scientific advancements to produce superior products. not only did the totality of the steps to manufacture a product become increasingly complex, but the work also was spread across departments. New supervisors were added to oversee these departments. And these new departments were further bundled and assigned a leader to supervise the departmental managers. This lead to the formation of a class of middle managers—a level sandwiched between senior leaders and departmental managers. From an organizational perspective, the result was a pyramid-shaped hierarchical leadership structure that was built primarily to supervise workers. Even today, this structure remains the predominant organizational structure in businesses and other organizational forms around the globe.

But a multitude of deficiencies has driven many business pundits to question the utility of the traditional organizational structure and begin exploring alternatives. The most prevalent issue is the flow of work. As work products are built, they move from specialist to specialist, crossing organizational boundaries until they arrive at the end consumer. The work is disconnected, performed in different locations by functional teams, and overseen by different managers. Quite frequently there are communication gaps or bottlenecks that occur where work products are transferred from one team to another. The priorities across the groups often differ, leading to competing and sometimes contradictory directives that further complicate the interactions between teams. The negative ramifications when disparate teams fail to work collaboratively led to the coining of the term *organizational silos.* From an innovation standpoint, organizational silos hamper efforts to collaborate across teams—effectively building walls between parts of the organization. A result of such conditions is that implementing any improvement is prohibitively difficult. Designing and implementing an end-to-end process requires the involvement and approval of any number of organizational leaders—who may well be motivated to different ends. Collaboration and unified planning are spotty at best and non-existent in many cases.

A slightly lesser known cousin of the organizational silo is what I label a *knowledge chasm*. Knowledge chasms represent the break in information flows between leaders and their direct reports. In essence, it is the professional workplace's version of the telephone game taught to us in our childhood. A message is whispered from child to child on down a line until it reaches the last one, who shares the version he or she received. On most occasions, the end message bears only a slight resemblance to the original. In the business world, variations of the telephone game impede the flow of information and prevent important details from arriving where they are needed. This occurs for any number of reasons. A frequently heard statement today in the corridors of corporate headquarters is that leaders don't need to get into the weeds. In other words, the details should be left to those who are closer to the subject matter. As a result, leaders are rarely exposed to the full picture of what is happening at ground level. Equally damaging, when information should flow to leaders from the field, overly ambitious managers censure what their supervisors hear—acting as gatekeepers to ensure that the presentation of the information is to their benefit. Lacking a smooth flow of accurate information, leaders debate opportunities and make decisions without a thorough understanding of the reality of the situation.

Silos and knowledge chasms are formidable roadblocks to innovation. If we return to the two ingredients of innovation—the need for a champion with the awareness and authority to fully engage the enterprise and accurate information on what the customer wants and what the enterprise can produce—it is immediately evident that there are major shortcomings in the structures and practices born of Adam Smith's theories. Who is the champion when responsibility is diffused across business groups? Who understands both the big picture and the details of how things get done on the front lines? How is change coordinated across the individually managed sections of a process? In entrepreneurial ventures, these innovation deficiencies are mitigated by the proximity of leaders to the front lines. They see firsthand what occurs at ground level. The organizational boundaries are blurred between sales, manufacturing, distribution, and customer service in smaller entities. This fosters a vibrant connectivity between leaders and employees that simplifies the flow of information and the execution of strategies. This leads many business pundits to suggest that large enterprises need to act more entrepreneurial. With growth, however, everything changes. New locations are added, software is implemented to manage information, employees are hired to fulfill sales orders, and product lines are expanded to further grow the market. The size and complexity of the enterprise increases with this growth. Organizational silos are birthed. Knowledge chasms begin appearing. The enterprise becomes less efficient and less responsive to market shifts. Unless the innovation roadblocks are removed, market share and financial performance are likely to erode.

CAN INNOVATION BE SYSTEMATIZED AND MADE REPEATABLE?

My first job after my college graduation was with the Fifth Third Bank in Cincinnati, Ohio. The economy was in a down cycle, and although I was lucky to have a job, I was not overly enthused about starting my career in the banking industry—especially at a regional bank in the Midwest that was not what many viewed as a real player in the financial industry. My enthusiasm took a further hit when I arrived at the corporate headquarters. Fifth Third prided itself on its spendthrift culture. Employees worked long hours in poorly ventilated offices with dilapidated furniture and threadbare carpets. Every penny spent by the company was meticulously scrutinized. However during the two years that I spent at Fifth Third, I learned more about strategy and innovation than any classroom ever taught me.

Shortly after my arrival, I was asked to participate in a sales blitz. Participation was mandatory for all levels of employees: cashiers, branch managers, loan officers, operations associates, technologists, and individuals from every department, including C-level executives. The goals of a blitz were to drive awareness and sales by canvassing a specific, predetermined neighborhood. As we walked a route, our directions were to connect with existing customers and introduce the bank to prospective customers. At the conclusion of the day, the blitz team reassembled to discuss our findings and share what we had heard during the day. A list was formed that identified promising prospects for follow-up. Even more importantly was a second list, a list that documented customer feedback. This second list was pure magic. We learned what the customer liked about our products and services—and we learned what frustrated them. By just talking with our customers, we identified real opportunities to give greater service to our customers. The sales blitz furnished leaders, managers, clerks, and other employees with a view of the bank from the perspective of customers—it gave us a window into their world and allowed us to empathize with them. And because senior leaders from across the bank were involved, the opportunities did not get tossed onto desks with the flood of other reports. They were investigated, and when they were seen to be valid, many ultimately were acted on.

The results speak for themselves. Fifth Third today ranks in the top 20 largest banks in the United States and is consistently recognized as one of the top-performing and best-managed financial institutions in the nation. Whereas most of Fifth Third's contemporaries of comparable size were acquired in the 1990s, Fifth Third methodically grew its asset base from $8 billion in 1990 to well over $100 billion today. What allowed Fifth Third to achieve such growth? At the core of its success is a focus on the customer and its ability to use ideas unearthed during sales blitzes to direct the evolution of the bank's products and services. Although never explicitly stated as such during my tenure, Fifth Third effectively created an innovation system.

Over the last decade, my consulting work and experience as a business executive have driven me to craft an approach not only to systematize innovation but also to improve the overall manageability of an enterprise. My goals were straightforward—to facilitate customer awareness, enhance operational adaptability to market opportunities, continually reevaluate and improve the efficiency of the enterprise, and develop a governing model to ensure the optimal allocation of energy and resources. The real challenge was connecting all these unique aims into a model that was both intuitive and specific. As my innovation framework took form, it quickly became clear that *the language of innovation is process.*

The greater business community has long embraced process as a construct to improve product quality, push down costs, and address deficiencies in the performance of work. Total Quality Management, Six Sigma, Reengineering, and Lean are reputable process based approaches with adherents throughout the business world. More recently, Business Process Management (BPM) took precedence as a discipline to systematize programs to improve process performance. These days, just about every midsized to large company has used one or more of these tools to address performance issues.

As a concept to describe work, processes are without peer. And they are significantly more powerful than most business theorists believe. Although there is a nearly unanimous acknowledgment of the potential of using processes to address efficiency opportunities, only rarely are processes recognized as the embodiment of an enterprise's strategy. But this is exactly what processes are—the unique way an enterprise consistently differentiates its offerings from those of its competition. Moving past their facade, one finds that every enterprise is a web of connected processes—a series of highways where work is performed until a final delivery is made to a willing customer.

This leads to a rather profound conclusion: process is a link between all types of innovation activities—a common language for communicating both strategic and efficiency adjustments. In contrast to its historical usage, process improvement is not simply a discipline to execute a one-time improvement. Processes are enterprise assets. We can assign ownership to them, guide their change, and use them as the basis for allocating resources where they contribute the greatest value. An enterprise using processes as the framework to manage performance and improvement efforts is a *process-focused enterprise*.

THE PROCESS-FOCUSED ENTERPRISE

The process-focused enterprise is the future of how enterprises of all types will be organized and managed. Gone are the organizational silos, gone is the command and control approach to managing people, and gone are all the guesswork and information inadequacies that plague the planning of an enterprise's future direction. Replacing such dysfunctional elements is an intuitive, simplified, fact-based, customer-connected, efficient approach to managing work activities. The core of this approach is the concept of process ownership.

To put it succinctly, *process ownership* is the assignment of an individual (or a group) to own an end-to-end process. In this arrangement, a process owner is responsible for the active management of all facets of a process. In this role, his or her responsibilities include

- Managing the performance of an end-to-end process on a daily basis.
- Adjusting the process to support strategic and operational improvement goals in collaboration with other process owners.
- Continually seeking methods to make the process more efficient without harming—or in concert with—business partners.
- Acting as a representative of the process and being able to speak to all its component parts, including resources, costs, ongoing and future improvements, metrics, and so on.
- Training and mentoring process performers.
- Understanding and speaking to the resource requirements of the owned process.

In theory, process ownership connects leaders with both the authority and the knowledge to get things done at the ground level. However, whereas operational awareness and ownership are significant steps forward in eliminating silos and engaging the full power of a workforce, they do not fully eliminate the risk of individual process owners working at cross-purposes, nor do they prevent the ambitious manager from operating in his or her own best interests. Taking enterprise performance to the next level entails aligning the ranks of process owners to a common set of strategic and operational initiatives—deploying an improvement schedule written in the language of process to confront the challenges of the day and prepare for tomorrow.

Such a system—such an approach—is the topic of this book. *Customer Focused Process Innovation* is an instruction manual to fundamentally empower leaders and guide them down the path to supercharge the ability of their enterprises to innovate. The goal is to bring new products to market faster, to operate daily with a customer-focused mentality, to rapidly adapt to new market conditions, to perpetually find ways to do more with less, and to capture market share by exceeding customer expectations.

2

The Innovation Recipe: Surfing the Wave of Change

As history clearly demonstrates, consistent performance in the business world is a rarity—achieved by only a scant few enterprises. But the fact that even a few are able to deliver year after year begs the question, "What are they doing different?" Despite a multitude of research studies and long-winded hypotheses, the code remains to be cracked. In fact, several prominent business theorists recently suggested that performance is more luck than anything. From my perspective, such a fatalistic tone ignores those successful enterprises that consistently delight customers, seize market share, and deliver bottom-line results year after year. Someone wins the battle in every marketplace. The question is, How are they doing it? The holy grail in the business world is to discover that elusive recipe for performance.

Why is it so hard to identify the levers to pull and the buttons to push to systematize success? One theory is that the business world lacks a sufficient number of cycles to appropriately identify an organizational model or approach that performs regardless of market and customer shifts. Change simply occurs too rapidly with too many variables—denying academics and the business community the time required to identify patterns and design models and techniques. And we know, there are only a handful of consistently successful enterprises to dissect and uncover what differentiates them from the rest of the pack.

To find the answer, perhaps our search needs to expand to investigate fields parallel to the business world with a greater track record of observations. The rationale behind identifying a comparative field is to study adaptive mechanisms—specifically those that enable organisms to thrive in threatening conditions. Fortunately, there is at least one such parallel—one that has been subjected to extensive research and analysis to arrive at conclusions that are both relevant and valuable to the business world—the natural evolution of species. The species populating our world, both today and in the past, and their tendency to flourish or to perish are analogous to market competition and the life cycle of enterprises.

Natural evolution is founded on theories developed in the mid-1800s by Charles Darwin and Alfred Wallace. Independently, but contemporaneously, they discovered that organisms evolve and adapt to changing environmental conditions through a process called *natural selection*. Depending on external stresses to a population, genetic variations allow a subset of the population to thrive under harsh environmental conditions. Other members of the population suffer or even perish under the same conditions. The surviving subset of the population eventually reproduces and passes the genetic advantages to their progeny. Over successive generations, those characteristics advantageous to the continuation of the species become increasingly prevalent in the greater population. In this way, species evolve and overcome detrimental conditions. The pace of natural evolution is modest and slow because it depends on generational adjustment.

The late Bruce Henderson, who founded the Boston Consulting Group and contributed much to the field of strategic planning, first identified this parallel between change in the business and the natural world. In his article, "Strategic and Natural Competition,"[1] Henderson states that understanding natural competition is a requisite to understanding strategic competition. Although he wrote of this linkage in 1980, when the field of strategic planning was relatively nascent, Henderson noted that evolution and strategic competition were nearly identical with one significant exception—time is compressed in competitive markets.

In contrast to the slow pace of natural evolution, strategic competition in the business world does not afford sufficient time to allow for the generational inheritance of advantageous characteristics. Competitive markets are in unending flux. Enterprise leaders are forced to respond to changing consumer preferences, aggressive competitors, and a host of other forces including societal, governmental, regulatory, and technological and scientific advances or else face the prospect of losing customers to more strategically attuned competitors.

To survive and prosper in the pressure cooker that is a competitive market, an enterprise must essentially replicate the evolutionary activities of a living organism—but at a much faster pace. For example, imagine that a new technology with far-reaching implications is released in the marketplace. The astute competitor mobilizes its resources to evaluate the implications and determines the appropriate course of action. Inside this company, a customer research team studies the technology's impacts from the customer's perspective and determines what the customer likes and dislikes. Elsewhere inside the company, a strategic planning team uses this research to plot potential market opportunities and predict how the competition might respond. The results of these assessments are passed to a leadership team that debates the merits of each opportunity and eventually selects a plan of action. The plan is implemented—and assuming that it was well designed and that the foundational assumptions behind the solution prove to be true, the enterprise's chances of success are vastly increased. The path undertaken by the company in this example is what I call the *innovation cycle*.

But innovation is never as simple as this example suggests. If leaders were able to focus their attention exclusively on innovation, they would be far more proficient at identifying market shifts and responding appropriately. But this is rarely the case. During an innovation cycle, customers continue to buy products, and all the work to deliver those products must continue. Similar to that ringing phone in the background, there are always distractions and competing priorities that demand attention. But even if the background noise were eliminated, innovation by itself is a complicated and challenging endeavor—with many moving parts. It is easy to back burner innovation planning as it is eclipsed by the cacophony of business chatter and demands for immediacy that come with running a business. This is exactly why having an approach to innovation is a necessity. And to ensure that innovation is not limited to a singular instance, the innovation cycle needs to be stitched into the very fabric of the enterprise and become an inseparable part of business as usual—exactly as it does in a living organism. In any enterprise, continual innovation depends on the institutionalization of adaptive capabilities and the existence of an organizational and management structure that supports reinvention. Such an approach needs to be methodical and scientific and pervade every corner of the enterprise. It must live in the leadership, culture, processes, organizational structure, and indeed the very DNA of an enterprise.

THE FOUR FACETS OF INNOVATIVE ENTERPRISES

Enterprises with a demonstrated ability to harness their resources and innovate their processes, people, and structures invariably embrace several key practices—practices that when performed collectively propel innovation. I call these practices the *four facets of an innovative enterprise* (Figure 2.1). The four facets are *customer focus*, *strategic planning*, *operational improvement*, and *initiative management*.

A note of caution, the mere existence of the four facets in an enterprise does not guarantee long-term or even short-term success. Prosperity depends on a multiplicity of factors—many outside the enterprise's sphere of control and others of sufficient complexity to thwart any accurate prediction of their impact. However, the facets are integral components of an innovative enterprise that must be understood and in existence at some level in order for leaders to appropriately respond to external change and internal challenges.

Here is how the facets connect together in the innovation cycle. Understanding what is important to the consumer (customer-focus facet) is the first part of the innovation cycle. Every enterprise intends

FIGURE 2.1 Four facets of an innovative enterprise.

4 Facets of an Innovative Enterprise

Customer Focus
Commits to understanding the motivations, behaviors, preferences, and processes of existing and potential customers

Strategic Planning
Identifies strategic initiatives after considering the customer, potential competitor responses, and the capabilities of the enterprise

Operational Improvement
Identifies operational improvement initiatives by analyzing the entire process system for cost reduction, throughput, capacity, quality, and safety improvements

Initiative Management
Incorporates strategic and operational efficiency initiatives into a coordinated and managed plan to generate the highest potential value for the enterprise

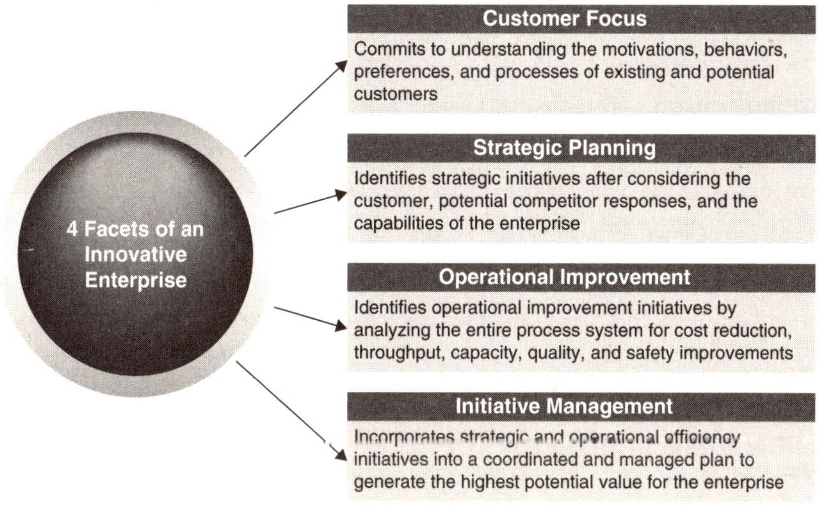

to create offerings that appeal to the consumer both today and, ideally, for some time in the future. Through customer analysis and with consideration of competitors' potential reactions, the leadership team identifies strategic initiatives (strategic-planning facet) to competitively position the enterprise's offerings in the marketplace. On an ongoing basis, the enterprise adjusts its productive capabilities (operational-improvement facet) to support the enterprise's strategic goals. Depending on the existing systems, facilities, equipment, and processes, this may include increasing the quality of products/ services, cutting customer response time, reducing operational costs, or increasing the enterprise's flexibility to respond to new opportunities. Whereas strategic planning is externally focused based on internal capabilities, operational improvement is focused internally with the goal of improving these same capabilities. The final facet (initiative management) is the collection, prioritization, and execution of both strategic and operational improvement initiatives. Simply knowing what needs to be done is only a part of the battle. An innovative enterprise aims to complete initiatives as quickly and efficiently as possible in order to seize the advantage before the competition knows what hit them. This is accomplished by prioritizing the full slate of initiatives based upon the value they are forecasted to deliver to the enterprise while accounting for dependencies and resource constraints. When this script is followed, the value delivered to the enterprise is maximized. Initiative management is not an annual event, like many strategic planning exercises today, but an ongoing activity that assimilates new information into the innovation cycle as it comes to light.

Taken collectively, the four facets are the innovation cycle for an enterprise. Each of the facets is much more than a simple checkbox on a planning template. To harness their potential requires a significantly deeper understanding of each of the facets and the critical activities comprising each.

FACET 1: CUSTOMER FOCUS

Customers are the reason for the existence of any business. As they collectively chose what products and services to buy, customers determine the winners and losers in a competitive market. This alone makes customer focus arguably the most important facet in the innovation cycle. Despite its importance, it is simply dumbfounding how few companies are attuned to their customers and invest the time and resources to understand how their customers shop, purchase, use, service, dispose of, and replace their offerings. Any enterprise that consistently ignores its customers will eventually lose their loyalty and their dollars to products more closely aligned with the customers' preferences. If deficiencies are not addressed in a timely fashion, the enterprise's continued existence may even be at risk. Unfortunately, the customer's voice is often relegated to background noise as more "pressing" matters steal leadership's attention.

In the business world, there is one definitive certainty. Customers will change over time—and not just one aspect of customers but every aspect. How they use the product, how they want it delivered, and their expectations of the product—they all change with the passage of time. This customer flux offers a major opportunity for the agile enterprise that sets the customer as its navigational beacon, but proves to be deadly to the lethargic competitor. To claim a share of the customer's wallet and remain relevant, an enterprise must reinvigorate its product and service offerings to match new customer preferences. The real challenge is to know what to change. Here is where many enterprises drop off the radar.

Across corporate America, management teams are more than happy to pay big dollars to consultants and research firms to gain insights into their customers. Why anyone would pay big bucks for information that is usually readily available and can be obtained largely free of charge is beyond my comprehension. What every enterprise really needs to be innovative is not reams of data and analytics but rather the *voice of the customer*. In other words, an enterprise needs to understand the customer's perspective on value—what is valuable or meaningful to them. From my experience, capturing the voice of the customer can be accomplished via a three-step process.

1. Capture and consolidate information on the customer from a variety of sources.
2. Analyze the data and build the enterprise's voice of the customer for every product/service family serviced by the enterprise.
3. Disseminate and make the voice of the customer available to the greater organization.

Although the steps seem relatively straightforward, they are commonly skipped, ignored, misunderstood, or not acted on. Without solid customer information on which to predicate business decisions, leadership teams shoot blindly in the dark.

Capturing Customer Information

Getting back to the availability of customer information, where does an enterprise go to obtain good, factual customer information? The short answer is that no single source suffices. In fact, the optimal method is to use a blanket approach—using varied sources as well as techniques for collection. Aggregating the information pulled from a variety of sources minimizes most biases and results in the most accurate picture of consumers' buying patterns and preferences. Diversity and the depth of the sources are directly correlated with the information's accuracy and potential to be actionable. From my experience, most companies already subscribe to information services or possess vast amounts of customer information internally. The real work is in locating, consolidating, and analyzing that information. Although there are many worthwhile sources of customer data, I find the following sources to be particularly valuable:

- Internal customer analytics
- Feedback loops and direct observation
- External customer research
- Trend analysis

Internal Customer Analytics

Perhaps the most commonly available source of customer information is internal customer analytics. *Analytics* is the capture and segmentation of data to learn more about the behaviors of customer subsets. This technique uses traceable data (including cash register receipts, loyalty clubs, and credit cards) to track the buying patterns of consumers. By doing so, the enterprise can analyze consumer purchases both at a singular point of time (i.e., a basket of goods) and for a set duration (e.g., annual purchases). Typical data gathered from this approach includes

- *Frequency of customer purchases or visits.* How often customers buy the product/service or visit a shopping location, such as a retail location or website.
- *Customer shopping basket.* What customers buy during individual shopping trips.
- *Product/service affinity.* The propensity of customers to buy certain products or services at the same time—for example, complementary or substitute goods.
- *Customer profile.* Identification of demographics and other characteristics to group customers with similar purchasing behaviors.

And this is only the tip of the iceberg. With expanding technology capabilities including the ability to track customers through loyalty programs and credit/debit-card purchases, customer data is easier than ever to obtain. In fact, the ease of gathering information creates a new challenge—consolidating and mining the information to pick out meaningful and actionable insights.

Feedback Loops

Although internal customer analytics provide useful insights into the past buying habits of customers, it is always beneficial to use additional techniques to corroborate knowledge gained from the cash register and enrich it with more predictive information. Arguably one of the best sources (although unfortunately one of the least used) is the capture of consumer information via feedback loops. No other technique provides such crisp qualitative information than reaching out directly to frontline associates and business partners and gathering their insights. As Sam Walton, founder of Walmart, once said, "The folks on the front lines—the ones who actually talk to the customer—are the only ones who actually know what's going on out there." Employees who interact with customers on a daily basis understand customer needs and desires at a greater level and with greater clarity than insights harvested by any research group. And the best part is that aside from the time to survey these employees, this information is absolutely free.

That said, there is a minor complication with this approach. Frontline employees may not be as easy to identify as in years past. In modern enterprises, quite a few positions may regularly interact with customers. For example, consider the historic role of research and development (R&D), where employees worked primarily with internal partners, such as the sales department, to develop new-product concepts. Today's R&D laboratory is radically different. Laboratory workers routinely partner with external customers and in some instances even collaborate with them (e.g., business to business) to build or redesign products. This complicates capturing feedback but in no way diminishes the number and quality of sources. To identify customer touch points, conduct a top to bottom review of the enterprise. Ask managers if their teams interact with customers on a routine basis, and if so, ask which individuals specifically. The identified names are the initial set of frontline associates to mine for feedback. However, time and collective mechanism limitations usually whittle down the list to the roles with the strongest customer connections. With these roles identified, put a communication mechanism in place to harvest feedback and customer insights from these employees on a regular basis. The options are nearly limitless, but the collection mechanism not only should be convenient to the employee but it should also support the free flow of information to a consolidation point. Mechanisms to capture feedback include surveys, information entry points on standard processes (e.g., entering competitive bids or the reason why a deal was lost), e-mail addresses to submit feedback, websites, periodic informational interviews, and focus groups. When gathered in a timely and methodical fashion, information collected from the front line is arguably the richest and most detailed customer information available. Whereas internal customer analytic provides raw data on purchasing behavior, feedback loops provide insight into the actual customer behind those numbers.

External Customer Research

External research is a good source of data to refute or confirm information gathered via internal mechanisms. The tools for capturing market and customer data include customer surveys, consumer research studies, market or industry research, focus groups, vendor/supplier studies, and a host of others. A significant reason to use external research when assessing customers is to counterbalance internally collected data with data from outside the walls (and the influence) of the enterprise. Whereas customer analytics and feedback loops focus on current customers, external research often spans both current and prospective customers, and this includes customers currently buying from the competition. External research can be bought for a specific topic or trend—and in many instances is available free from industry or government sources.

One caveat for external research: it is always appropriate to question the applicability and validity of the data. Resist collecting information simply because it is conveniently available. Gather information that is accurate and applicable to the marketplace. Remember that in most cases the information was not collected, cleansed, and interpreted solely for your enterprise. It is generalized and often simplified information.

Trend Analysis

A final source of customer information is trend analysis. While not calibrated directly to a current or prospective customer, this information focuses on larger trends that influence overall customer preferences and buying behaviors. Trend analysis casts a wide net and considers not only the trends that potentially affect the immediate market (micro trends) but also larger, more universal (macro) trends. Macro trends include the impact of change from various angles, including technological, societal, environmental, political, regulatory, demographic, attitudinal, and behavioral. Although such a wide swath may not generate opportunities that are immediately actionable, it only makes sense to plan customer solutions with consideration of the macro and micro trends.

Analyzing Customer Information and Predicting the Future Customer Profile

Although many enterprises collect information about their customers, the benefit they glean from that information is only partially realized because they fail to complete the critical step of making the information actionable. What they are missing is the activity to aggregate the information and pass it to strategy and planning teams. Without this step, the information is just a bunch of data hidden away in the corporate data warehouse, sitting around on computer drives, or left on paper reports scattered about the office.

The aim of customer analysis is to open a window into the mind of the consumer. However, before a vivid view of the customer can be created, the information needs to be consolidated into a single repository and sliced and diced to derive meaningful insights. For example, rarely does an enterprise offer only one product. Therefore, one of the first actions is to segment the data by product family or other market differentiator. Here a bit of common sense must be infused into the process. Product/services families are a distinct offering sold to a group of customers. In many instances, there are synergies between product/service families. In fact, identifying synergistic products and bundling them as an offering is a strategic option. In a similar manner, the information may be sliced to identify the products/services sold to specific types of customers, such as distributors. Initially, segment the data consistent with the enterprise's go-to-market approach. After that, a good analyst will search for other meaningful segmentations. If the segmentation does not prove meaningful and actionable, discard it and keep going.

After segmentation, the data is ready for further analysis. Kano analysis is one of my favorite tools for mapping customer preferences. This model (Figure 2.2) summarizes the customer's preferences by identifying attributes of a product/service offering and the market's perception of that attribute.

The product/service attributes are charted based on customer satisfaction (or dissatisfaction) and whether the attribute performs well or not. In the Kano model, the categories include delighters, performance needs, basic needs, and dissatisfiers.

FIGURE 2.2   Kano analysis.

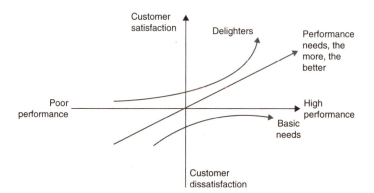

- *Delighters.* Delighters are attributes that are unexpected but appreciated by customers. They provide the "Wow!" factor that motivates customers to tell their friends about their experience. Delighters are not always immediately apparent to customers—often they surface during continued use of the product/service.
- *Performance needs.* Performance attributes are characteristics of the product/service that are the prime determinant of whether the product provides value. In general, more is better because the product exceeds what was anticipated or communicated during the shopping and purchasing processes.
- *Basic needs.* These are the attributes of the output that are expected by the customer. When basic needs are not a part of the output, customers will shop elsewhere. If they are included but do not perform, customers become dissatisfied and will, at a minimum, shop more broadly in the future.
- *Dissatisfiers.* These are attributes of the output that are unwanted by consumers. If they are known before purchase, customers will likely pass up the purchase.

An equally important aspect of this analysis is the *latent attribute*. Latent attributes are unknown by the customer. They are similar to delighters in that their discovery may drive satisfaction, but they differ in that they meet a need or want that the customer was not aware of during the purchasing process. The end goal of Kano analysis is to understand what the enterprise needs to deliver to increase sales and customer satisfaction. In Kano terms, this is accomplished by meeting the basic needs of the customer, performing to the customer's expectations, eliminating dissatisfiers, and delivering a healthy dose of delighters.

Kano analysis is far from the only method to aggregate customer requirements. An increasingly popular approach is to use personas to model customer behavior. *Personas* or *virtual customers* are profiles built to represent the differing perspectives and characteristics of a specific customer segment. By modeling a set of customers via a persona, an enterprise can predict how different groups of customers might react to product enhancements and marketing programs.

Once the customer information is aggregated, segmented, and digested, many leaders want more predictive insight—especially with regard to the direction and momentum of customer change. What will the customer want in the future? Obviously, awareness of the trends affecting consumers and the overall market improves the quality of the prediction. However, at the end of the day, the best answers are derived from intuitive guesses based on available information. Making the customer data widely available fosters additional debate and discussions and helps to forge a collective mind-set on the evolving customer and what might be critical to satisfying future customers.

Distributing the Customer Profile

The output of the customer information-gathering process is a multi-dimensional view of the customer—a view that can be widely distributed to support more intelligent strategic and operational planning.

Although this information aids good decision making, leaders should remember that the customer is always evolving and that every idea requires a level of testing prior to release. Research only guides the discussion—it does not provide the answer. Even if perfect information were available, the charted strategic course is only a best guess. It almost always takes a number of iterations to accurately pinpoint a promising market position. The process that takes informational inputs and develops this plan of action is the *strategic planning process*—the next facet of an innovative enterprise.

FACET 2: STRATEGIC PLANNING

Strategy and *strategic planning* are two of the more overused and least understood terms in today's business vernacular. The word *strategy* pervades business conversations and has come to refer to any planning activity. Today you hear managers talk about having an information technology (IT) strategy, a strategy for completing a project, and even a strategy for improving the efficiency of a process. Unfortunately, the term's overuse has led to a widespread misunderstanding of critical activities for plotting an enterprise's future. For the sake of clarity, strategy and strategic planning are externally focused and refer to activities that competitively position a company in a marketplace. The intent of a strategy is to grow market share and ideally to do so in profitable markets. Strategy encompasses game theory—where a company positions its offerings to gain advantage against competing offerings. It predicts how a competitor will react and therefore allows a company to understand the true financial potential of strategies. In a market without competitors, strategic planning is unnecessary. In the for-profit world such a situation only exists in the most managed and regulated markets. Once a company decides on its strategic intentions, the collection of activities to position the company in alignment with these intentions is a *business plan*. The litmus test to determine if an action is strategic is to ask, "Is this activity directly aimed at increasing market share?"

Today, strategic planning is perceived as high profile—an esoteric practice in which only the elite and most influential leaders are given a seat at the table. And here begins the trouble with contemporary strategic exercises. The reality is that large personalities dominate the process—often promoting goals to enlarge their sphere of influence. Not uncommonly, employees with valuable insights or ground-level knowledge are excluded from participating in the strategy-formulation process. And in the absence of strategic expertise, strategies are conveniently copied from the competition. After all, it is easier to mimic the actions of others than to generate new, risky ideas. Although imitation may be the sincerest form of flattery, it is a horrible way to innovate. Unfortunately, this is the path many leaders chose to follow.

In the most recent decade, business theorists and thought leaders pushed metric-based goals to the forefront of strategic planning. From a managerial perspective, the prevailing opinion on metrics is that what gets tracked gets done. Although there is some credence to this theory, metrics are limited in their capability to link strategic intent with ground-level execution. However, once the metric drug took hold, enterprises began tracking all sorts of stuff and generating reams of reports. It soon followed that leaders expected that the numbers could be analyzed and used to predict the optimal course of action for an enterprise to follow. Such an approach conveniently ignores the often-quoted statement on glossy investment dossiers: "Past results are not indicative of future results." Because of the dearth of strategic planning know-how in the leadership ranks of corporate America, analytics are receiving a tremendous amount of attention from executives today. Rather than making a decision on the strategic direction of the enterprise, executives want the numbers to make the decision for them. But it does not work that way.

Analytics, in my opinion, are misapplied when leaders look for forecasts to guide their actions. The future cannot be predicted—even using the most comprehensive knowledge available and the latest technology. Were it possible, business and investment analysts of all stripes and sizes would immediately lose their jobs because their opinions would be supplanted by predictive facts. However, although analytics cannot provide *the* answer, they are useful as the background on which to predicate strategic exploration.

Strategic planning—indeed, any future planning—requires intuitive decision making. Intuition is not the use of divining rods. It starts with a healthy collection of unbiased information on the existing market environment, understanding the momentum of the forces at play, and then generating solid intuitive hunches on the most promising opportunities. It is this "sense" that drives entrepreneurs to take a second mortgage and to start a new company. It is generating this intuition—a hypothesis based on an abundance of information culminating in a gut feeling—that is the aim of the strategist.

This leads us to one of the greatest struggles with strategic planning—assessing its effectiveness. The lag between strategic actions and market results makes correlating the action with the result essentially guesswork. Although a strategy might be spot on, the benefits may not manifest for years. In a period when investors clamor for results, leaders do not have the luxury of waiting. And further complicating the analysis of strategies is the question of whether a specific strategy failed because it was misdirected or because of poor execution. But these difficulties should not dissuade leaders from advocating for a robust strategic planning function. Strategic planning vastly increases the odds that an enterprise responds correctly to shifting market conditions and is an infinitely more effective approach than letting the chips fall where they may.

But what makes a strategic planning process good? What are the key activities in a good strategic planning process? To start, let us review a list of activities included in a best-in-class strategic planning process. Rarely today does an enterprise have the knowledge, structure, or processes to execute all the activities I have listed. From my experience, customer analysis, competitor analysis, and game theory are the steps most frequently neglected or skipped.

Activities in a World-Class Strategic Planning Function

- Identification of market and customer segments in which the enterprise competes
- Detailed view of customer wants and needs
- Identification of competitors and their market positions
- Analytical and intuitive identification of potential market offerings (i.e., product, fit, positioning, etc.)
- Understanding of the enterprise's capabilities relative to the strategic alternatives
- Development of initial strategic hypotheses
- Use of game theory to predict the market reaction to strategic options
- Selection of optimal results based on competitive market models
- Development of objectives to accomplish the desired strategic result
- Creation of strategic initiatives to accomplish those objectives

To be effective, these steps need to be more than just formalized checkboxes. Building a good strategic plan requires diligence in gathering solid, unbiased information and processing that information to identify opportunities. Thus any good strategic planning cycle begins with an all-out effort to collect a solid book of information on both customers and competitors for a specific product/service market.

In addition to customer information, strategic planning requires details on competitors in each product/service market. This includes information from historical and current perspectives on how competitors position their products/services. In other words, what is the value proposition that the competition communicates to customers? Fortunately, much of this information is readily available.

Using the information gathered about customers, the strategists identify opportunities to expand the sales of existing products/services or to develop new products/services. That is, they find places in the market where they can exploit their relative advantages to steal market share. In a world of imperfect information, strategy and market positioning come down to intuition—a hypothesis as to what customers value and how the company can uniquely provide it. With the identification of a strategic hypothesis, game theory then comes into play. A strategic planning team studies the competition to predict the reaction to the company's strategic hypothesis and the possible financial outcomes of the strategy's implementation. The aim is to identify strategies to maximize sales and profitability while preserving the overall profitability in the market.

Preserving the profitability of a market is an important yet often overlooked aspect of strategic planning. When companies ignore the competition's reactions, they make themselves and their market vulnerable to a major upheaval. For example, a company that opts to pursue a cost-reduction strategy by dramatically reducing its retail prices may well gain market share initially. However, if the competition chooses to respond by also cutting prices, the overall profitability of the industry declines. Commodity products are particularly susceptible to price wars. After a protracted price war, raising prices and the return of the industry to a more natural level of profitability may be extremely difficult, if not impossible. Astute leaders avoid price wars and other destructive strategies because of their potential negative ramifications.

To understand a market, I find it helpful to graphically depict the customer/provider information into a user-friendly format. One useful tool to depict markets is a *market map*. A market map (Figure 2.3) is a two-dimensional diagram that identifies the customers on the vertical axis and the suppliers on the horizontal axis. Where a specific customer intersects with a specific supplier, the dollar amount of the sales for the product/service is noted. To see patterns, it helps to represent the sales graphically through the use of shading (denoting the size of the relationship) or via icons (with each individual icon equating to a set amount of sales). Along the borders opposite the customer and supplier listings, the specific attributes of the supplier or customer are listed. At a glance, the market map quickly identifies rough market share and the product/service attributes desired by

FIGURE 2.3 Market map.

	Suppliers				
	Strategy Firm	Supply Chain Consulting Firm	Specialized Consulting Firm	General Consulting Firm	Customer Summary
Project Resources	🚶🚶🚶🚶🚶 🚶🚶🚶🚶🚶			🚶🚶🚶🚶	• Projects support corporate initiatives • Few Buyers
Staff Augmentation		🚶			• Staff Augmentation roles to support IT PMO • Other staff aug roles Eliminated
Supply Chain	🚶🚶🚶🚶🚶 🚶🚶🚶🚶🚶	🚶🚶🚶🚶	🚶🚶🚶🚶🚶 🚶🚶🚶🚶		• Network Integration started with two assessments
Strategy	🚶				• Internal resources believed to be experts • Lingering bad taste from Strategy Firm Failure
Provider Summary	Revenue = $1050K IT PMO, Shrink Initiative, Lean Assessment, Cost Reduction Initiative	Revenue = $250K Supply Chain Assessment	Revenue = $500K Distribution Network Assessment	Revenue = $200K Category organizational Assessment	🚶 = $50K

Customers (Buyers) (vertical axis label)

the different customer segments. A market map may be segmented by other dimensions (e.g., geographic, temporal, or specific customer segment) for a deeper investigation of opportunities. Building the most useful market map may take several iterations. The aim is to define customers by what they buy, how they use the product, and what they value in a product.

Similar to customer profiles, building a current-state market map is only the first step to building a market map to predict a market's future. This is true strategic planning at its most fundamental level. Based on a projected future market view, strategists can identify potential product enhancements or where they can promote their existing offerings to new customer segments. A single market map rarely provides all the information critical to identifying strategic options. Large conglomerates with product portfolios spanning multiple industries require multiple maps. Strategists and market analysts often create multiple versions to examine variants of a single market. Again, a market map is not the end state of customer understanding but rather a tool to identify where to attack in a market—and what positions to defend.

Through the analysis of market maps, customer data and other market information, strategic opportunities surface. And when it comes to building strategies, no strategists need fret over a lack of advice. Bookstores sell literally troves of books offering guidance on how to build game-changing strategies. Before even getting to strategies, though, one precedent question helps to eliminate a lot of wasted effort. For each product/service family, ask if this is a market worthy of participation. If the profit is low or the risk is high, perhaps there are other more promising targets. When there is potential in a market, then proceed with the selection of a strategic course of action. From my experience, there are only four main strategies—although variants of each exist.

■ *Strategy 1: Low-cost provider.* The intent is to sell the product/service at the lowest possible price to consumers.

This strategy is extremely common, is often successful, and is relatively low risk for a market leader in an established market. The enterprise with the highest market share possesses an advantage when choosing this strategy because of the benefits associated with climbing the learning curve faster and gaining economies of scale.

- *Strategy 2: Innovator.* The enterprise choosing this strategy commits to continually offering products and services with new benefits and features of interest to consumers. This strategy is extremely risky but comes with a tremendous upside. A first mover enjoys the benefit of creating customer preferences and forcing the competition to react. The key for this strategy to be beneficial is to base the innovations on the unique advantages of the enterprise. That is, make them hard to copy. Otherwise, low-cost providers may swoop in and nab market share before the innovator has recouped its investment.

- *Strategy 3: Value player.* Enterprises following this strategy seek to provide the greatest value per dollar spent by consumers. While not a low-cost provider, such enterprises provide product/service attributes exceeding those of the low-cost provider and at a price commensurate with this additional value. Differentiators frequently include an expanded suite of offerings, reputation, durability, flexibility, quality, efficacy, and others.

- *Strategy 4: Customized product/service provider.* The last strategy focuses on providing individualized products/ services for consumers. This strategy is differentiated by the development of a product or service specifically tailored to specific customers. Luxury goods fall into this strategy, as do highly customizable offerings such as high-end automobiles or yachts. As advancements in manufacturing and customer communications progress, the cost of and obstacles to customize for a larger audience are falling—making it possible for manufacturers to offer individualized products such as shoes at Nike.com.

Strategy is more than identifying market opportunities. The product or experience actually has to be delivered and accepted by the all-powerful customer. Before a strategy is rolled out, there always should be a confirmation that it aligns with the enterprise's capabilities. This includes a study of the value chain and its adaptability to manufacture the desired products/services or whether the capabilities need to be built from scratch. Equally important is whether the strategy is credible to target customers. Will they embrace it as a logical extension of the company's offerings, or is it tantamount to Burger King opening a gourmet restaurant? Smaller companies often skip this step because the ramifications of a promotional misstep are minor. They lack the brand image to damage. However, as larger enterprises extend their reach into new market segments, they often try to be all things to consumers. Such a strategic move may backfire with customers who have grown accustomed to the company's offerings and brand message—destroying any goodwill previously earned and pushing consumers to consider alternatives. Strategy is as much about deciding what a company will not do as it is determining what a company will do.

Once a strategic objective is selected and its feasibility is confirmed, the strategy is broken down into pieces called *initiatives*. Strategic initiatives are the culmination of the strategic planning process—taking high-level opportunities and separating them into manageable chunks of work. For every identified strategic objective, there will be one or more strategic initiatives. For example, a company may key on a strategic objective to expand its existing product offerings. A strategic initiative to support this objective may be to launch a new product line for a newly identified customer base.

Another initiative may be to adjust an existing product in response to changing consumer preferences. These initiatives focus on different customer segments and products and therefore require two distinct efforts, but both initiatives support the same strategic objective of expanding the company's offerings.

As a simplification, strategic initiatives provide the "What?" and the "How?" of competitive positioning. But even with a perfect strategy, market success is in doubt if the enterprise is unable to deliver. Strategic execution relies on the existence of internal capabilities to produce the desired output. This brings us to the third facet of an innovative enterprise—*operational improvement.*

FACET 3: OPERATIONAL IMPROVEMENT

If the intent of strategic planning is to drive to a specific destination in a car, the goal of operational improvement is to keep the engine and other systems operating efficiently so that the destination is attainable. In a nutshell, operational improvement encompass the development and ongoing enrichment of an enterprise's operational capabilities. Its scope includes internal functions, processes, and resources that sum together to make it possible for a company to design, develop, manufacture, sell, and deliver its offerings. When viewed from a strategic lens, operational improvement provides maneuverability, adaptability, and flexibility—delivering the ability for an enterprise to chase sales opportunities and to do so efficiently. Measures of operational improvement include profitability, scalability, and adaptability.

■ *Profitability* measures the efficiency of operations to produce customer-desired outputs. In the accounting world, it is measured by subtracting total costs from total sales. Operational improvement focuses on the cost side of this equation—delivering units of output at the lowest possible cost.

- *Scalability* is the ability to adjust production to support volume increases or decreases efficiently. Throughput is a measure of the amount of output produced in a given period of time.
- *Adaptability* is the ability to transition production to new products/services or geographic markets. Adaptability is a measure of strategic maneuverability of an enterprise's structures and processes.

Unfortunately, operational improvement is undoubtedly the most neglected of the four facets. The old adage "If it ain't broke, don't fix it" is alive and well. In most enterprises, significant functions fall outside the domain of strategic initiatives and lack the "sexiness" to receive the attention they deserve. This is a colossal mistake. Supporting, managerial, and administrative (SGA) functions are frequently victims of such neglect. When considering that the cost of these functions may be the lion's share of overall corporate costs in some industries, SGA areas are some of the most fertile grounds for reaping big returns on improvement endeavors. An added benefit is that the savings recouped from efficiency projects flow directly to the bottom line. And from my past experiences, I know that operational initiatives often pack as much punch as strategic initiatives in moving the financial needle. During periods of market stagnation, the redirection of improvement efforts to operational areas may be particularly effective when there is little room for strategic maneuvering. I know of several companies that rely almost exclusively on efficiency programs to fund their investments in strategic endeavors.

Given the functional silos that exist in many enterprises, department-based efforts to elevate productivity are unfortunately the norm.

The results of these efforts are all too often counterproductive—improving a section of the value chain at the expense of the overall system. To a large extent, the Reengineering movement was the culmination of a near-universal failure of enterprises to execute major improvement efforts across organizational divisions. Until the mid-1990s, business managers attended to deficiencies in their areas without consideration of the greater operational impact. In contrast, the Reengineering movement aimed to create game-changing advantages by tossing out the old way of doing business and rebuilding the value-creation engine from the ground up. In this way, a company could leapfrog competitors who plodded along using outdated improvement approaches.

Although the primary focus of operational-improvement initiatives traditionally has been to improve efficiency, a complementary and equally important aim is the expansion of strategic capabilities. Especially in hypercompetitive markets, flexibility and adaptability are critical to success. Under most market conditions, price cutting supported by internal cost reduction is the safest strategy to undertake (and therefore also the most popular). Cost reduction improves a product's profitability and may extend its life cycle. Lower prices are always appealing, but the benefit may be short term if competitors retaliate by also lowering their prices. In contrast, changing a product's attributes comes with the risk that customers may not embrace the new design. The counter to a cost-reduction strategy occurs when new innovations make the (less costly) product obsolete.

In contrast to strategic initiatives, operational-improvement initiatives are generated in a different manner. Whereas strategies are birthed through the strategic planning process, operational-improvement initiatives capture attention in multiple ways, including the evaluation of quantitative or qualitative metrics, an investigation of cost pools inside the enterprise, the uncovering of "limiting" enterprise capabilities (e.g., systems, processes, or resource constraints that impede the enterprise from moving forward with other initiatives), and grassroots identification by employees.

With increasing frequency, enterprises dedicate teams to identify and capture operational-improvement opportunities. This may be done via a special-forces approach, where a designated team takes the mantle of a pseudo–efficiency squad and is set loose on the enterprise to find what it may. A close cousin to this approach is to conduct a health check on the enterprise through an enterprise-wide investigation to identify opportunities and bundle them into initiatives. Another more recent approach is to use a team of Lean coaches schooled in specialized efficiency tools to coach functional leaders and managers on opportunity identification. This option, while often slower, institutionalizes the capability to improve performance—and fosters a grassroots effort to exploit opportunities. When an enterprise reaches the point where managers are perpetually improving their respective areas, the advantages compound over time.

Regardless of the method of identification, the desired end state is a list of ideas to reduce cost, increase safety, expand capacity, or improve quality. From these ideas, initiatives are created to bundle the ideas into actionable pieces of work, exactly as done with strategic initiatives. These initiatives commonly take one of the following forms:

- *Process improvement.* Using tools such as Six Sigma (to improve the quality of a process), Lean (to remove waste from a process), and Reengineering (to transform the enterprise's processes and drive breakthrough results).
- *Technology enablement.* Automating processes or using information to improve how processes run (e.g., enterprise resource planning systems).

- *Organizational design.* Focusing on how people are organized to perform work, how the performers are trained for their role, and motivating them.
- *Structural improvement.* Investigating and improving the systems that are in place to support the enterprise production equation (e.g., the production system, compensation system, cultural behaviors, and leadership styles).

As stated previously, the overwhelming majority of companies in corporate America largely ignore operational improvement except when they inhibit the execution of corporate initiatives or during market downturns when costs are scrutinized. In contrast, innovative enterprises aggressively seek opportunities to improve processes, functions, and structures.

The strategic and operational facets are the ying and yang of innovation. Both are necessary for an enterprise to prosper year after year. Depending on economic or market conditions, though, one of the facets may take precedence because it promises greater rewards. It is not uncommon for the operational focus to take a back seat during a period of rapid market change as an enterprise fights for market share and operations are sufficient to the task. Correspondingly, a strategic focus may be less necessary during periods of market stagnation when operational improvements offer the greatest potential for increasing profitability. But periods of dominance for either facet are short-lived, and prosperity over an extended period of time requires that enterprises pay attention to both.

When strategic and operational initiatives are viewed collectively, the consolidated list is an *innovation portfolio*—the sum total of efforts to improve the strategic position and financial results of the enterprise. The management and execution of these initiatives are what drive continual innovation—and doing so efficiently and effectively is the goal of *initiative management*—the final facet of an innovative enterprise.

FACET 4: INITIATIVE MANAGEMENT

Three of the four facets focus on the path to identifying strategic and operational opportunities. In contrast, *initiative management* is the preparation, evaluation, staging, and execution of the initiatives to actualize those opportunities. The result of initiative-management activity is an innovation game plan—a list of initiatives prioritized in the order they will receive resources and be executed. Taking it to a deeper level, initiative management includes the scoping, prioritization, allocation of resources, launch support, and ongoing management of the complete collection of initiatives. Some people will argue that this function is performed adequately in corporate America today. Here I vehemently disagree.

Today, management of the portfolio of improvement activities is—with rare exceptions—vastly underdeveloped and poorly defined. Similar to support and governing processes, the initiative-management function languishes from inattention and clarity as to its purpose. Rarely, if ever, is it the target of a formal improvement effort. As a result, it hinders innovation efforts by focusing attention and resources in the wrong places.

Whereas it is by itself extremely rare that an enterprise prioritizes its initiatives, I know of only a few enterprises today that fully incorporate resource allocation into this exercise. Thus enterprises are, for the most part, investing blindly in their future. This is a mistake on many levels. The rationale for including all initiatives in the same launch process is to enable smarter investment decisions. Resources and energy should be focused on efforts predicted to deliver the greatest bang for the buck—that is, initiatives that deliver the greatest value. And when the process expedites delivery of the benefits, the value of the innovation portfolio is maximized.

Today the most typical initiative-management approach is to approve initiatives on the basis of their individual merit—in other words, on whether they have a positive impact on the bottom line. Each initiative is evaluated from a go/no-go proposition with minimal comparison with alternative investments. When given the green light, the initiative is tossed into the chute for execution—many times leaving planning and resource procurement to a sponsor or the eventual project team. Obviously, executing initiatives in isolation omits consideration of collaboration opportunities and dependencies. Additionally, this approach blatantly ignores the scarcity of resources. If an enterprise has unlimited resources and no competition, this approach is perfectly reasonable. However, most enterprises operate in markets with numerous competitors, and they have limited capital to invest, limited technological capacity, and limited qualified personnel to toss at improvement opportunities.

Although not the blatant outright victim of neglect like operational improvement, initiative management is arguably the worst-performed facet in contemporary enterprises. Many enterprises, especially larger companies, use portfolio-management organizations (PMOs) to oversee and track the execution of major projects. Although launched with the best intentions, PMOs usually degenerate into a pseudo–status-tracking mechanism. Project and program managers pull together reams of documentation, including charters, work plans, financial workbooks, stakeholder assessments, and various other "required" project deliverables. From then on, the project's documentation and status are reviewed at delivery stages (often called *gates*) or periodic intervals (e.g., quarterly). To quickly communicate their stage of completion, projects are assigned a color correlated with stoplight colors. Green denotes a project on target based on the original estimates of budget and timeline. Yellow denotes a project at risk of falling behind the original timeline or going over budget. Red identifies a project in trouble. What I find humorous is how "red" projects are commonly handled. When a project slides off the rails, a change order is created to adjust the deadlines or budget. Once approved, the project reverts back to a green status. The fundamentals remain unchanged. Is this really an efficient way to build an enterprise's future?

In addition, many initiative-management processes suffer from the absence of integral elements. Most prominently, they lack any true health check on the initiatives in progress. Reviews rarely extend beyond timelines and budgets and usually fail to question whether the initiative's goals are still relevant or even desirable, except in the most obvious situations. As a result, it is rare to find initiatives that are halted or canceled even when their potential for success is negligible. Along the same lines, initiatives in flight are rarely redesigned or course-corrected when the underlying assumptions change or their current path leads to nowhere. In effect, initial goals and project directives are gospel. After launch, there are no remedies for inaccurate assumptions or flawed planning.

Under these conditions, initiative management (or *portfolio management*, as it may be called) ends up being more of feel-good exercise than an actual evaluation and management process. And this is where failure takes root. The wrong initiatives are resourced and launched—and more strategically and financially beneficial initiatives are back-burnered. Although perhaps not initially visible, the enterprise gradually loses ground to more nimble and assiduous competitors. These missteps affect profitability and the strategic footing of the enterprise. This leads to an equally big issue: enterprises rarely, if ever, assess and improve their approach to managing and executing initiatives. In the absence of such an examination, initiative management remains the same checkbox activity.

Understanding the potential of good initiative management necessitates a good definition for *initiative*. In today's business world, initiatives are usually associated with strategic planning, and in this context, they are identified as a collection of activities to achieve a strategic objective. But let's broaden this definition slightly: initiatives are programs or projects, not necessarily strategic, that when collectively executed deliver a financial benefit or expand the enterprise's capabilities. As mentioned previously, consolidating all of an enterprise's initiatives into a single list yields the innovation portfolio.

The total value projected to be delivered by the innovation portfolio is the sum of the individual benefits from each approved initiative. But the total value of the innovation portfolio is not static. Indeed, it rises and falls with the value of individual initiatives. Because of the time value of money, an initiative that is delayed for a year will, all else being equal, have a lower net present value. The aim of portfolio management is to maximize the value generated by the innovation portfolio by expediting the highest-value initiatives and stopping or delaying initiatives with minimal or no benefit. Tools and techniques for maximizing the value of the innovation portfolio will be discussed at length in Chapter 7.

INNOVATING AN ENTERPRISE VIA THE FOUR FACETS

The four facets of innovative enterprises are the primary ingredients for systematically responding to an ever-changing world. That being said, circumstances rarely require an enterprise to be hitting on all cylinders at any one time. Enterprises go through periods when the cash register sings and strategic adjustments are delayed as the company strives to keep pace with incoming customer orders. Other times sales are slow and cost savings are needed to keep the business afloat. Enterprises employing an innovation approach consistent with the four facets are armed to react appropriately and prosper in both evolutionary and revolutionary markets.

Building a continuously innovative enterprise entails institutionalizing the facets by melding them into the DNA of the enterprise. This is where most improvement theories run into a wall—how to bake these critical practices into the everyday activities of an enterprise.

Based on my experiences, I believe that the optimal framework for innovation is to use common terminology and structures. Processes are the structure that most accurately depicts work efforts. By using processes as the basis on which to frame improvement activities, the facets can be seamlessly intertwined into the structure, people, and management of any enterprise. The first step on this path is to fully understand the concept of process.

3

The Power of Process

As taught in basic operations courses, the work efforts in any enterprise can be organized into processes—step-by-step ordered tasks that collectively perform a job. Ideally, value is created as processes are executed, but this is not always the case. On an enterprise level, the recipe is fairly straightforward: an enterprise obtains inputs, converts them through the performance of processes, and produces outputs such as finished goods, services, or information. When successful, the outputs appeal to consumers, who purchase the product/service. Because processes are the foundational construct of value creation, the way they are designed, organized, managed, and executed translates directly into an enterprise's performance. If leaders and managers want to elevate the performance of their enterprise, processes are the appropriate starting point because not only are processes the drivers of value creation today, but they also represent an enterprise's capability to produce value in the future. The importance of processes cannot be overstated. But what really are processes?

DEFINITION OF PROCESS

A host of differing opinions exist as to the correct definition of *process* and what constitutes a process. In the book, *The Agenda*, the late Michael Hammer [1] identifies processes as an organized group of related activities that work together to transform one or more inputs into outputs that are of value to a customer. But there are competing definitions as well. Peter Keen argues in his book, *The Process Edge* [2] that Hammer's definition is overly restrictive and that a process may in fact not have obvious inputs and outputs. Keen defines a process as having four criteria—it is recurrent; it affects some aspect of organizational capabilities; it can be accomplished in different ways that make a difference to the contribution it generates in terms of cost, value, service, or quality; and it involves coordination. And the debate rages. Nearly every consultant, academic, and business theorist conversant with the concept of process has an opinion.

As a field practitioner elbow deep in process-improvement efforts on a daily basis, I see that the definitions advocated by Hammer and Keen are unnecessarily limiting and overemphasize "clean" processes that are immediately visible and require minimal abstraction to understand. Their definitions contradict what I see in the real world. Across industries, most process are unclear, unmanaged, unorganized, and—all too often—valueless. Only in the most sterile of settings do processes consist of an organized group of activities that work together. A sizable proportion of processes are ad hoc in nature, lack any planned design, and all too frequently move in opposition to an enterprise's stated goals. Further, you would be hard pressed to find value in many processes—especially those created to comply with government and regulatory agency mandates. And to stipulate that to be labeled a process, a group of activities must be recurrent or require coordination is a view from a purely theoretical perspective. Customization and oversight processes are rarely recurrent, but they are integral elements of value creation. Having slandered the popular definitions sufficiently, I suggest a far simpler definition for process: *processes are activities that use inputs to produce outputs.*

To further clarify this statement, process inputs can be anything— raw materials, capital, employee time, equipment usage, methods, tools, or knowledge—anything. Likewise, outputs can be anything— a finished good, a document, knowledge, a service, a decision, or even the lack of a decision or finished good. Processes do not necessarily create value for an external customer or even an internal customer— in many cases they accomplish little or nothing, and at their worst, they destroy value.

THE ROLES PROCESSES PLAY

Although most frequently cited as a tool to organize work efforts, processes play a far greater role in the development and management of an enterprise. Processes serve six primary roles:

- *Guidelines for the daily execution of work.* The specific activities that workers complete on a routine basis to build products, services, and information. Processes are a necessity for consistent execution.
- *A framework for continual improvement.* A holistic view of the interconnected and interdependent activities that together encompass the work performed across an enterprise. Processes allow leaders to see how the pieces fit together and to plan improvements with consideration of their impact on the greater system.
- *A foundation on which to create and track metrics.* Identification of the increments of work to enable measurement and reporting.

- ■ *Tools for training.* Documentation of the sequential work steps to deliver an output; this documentation is used to educate workers and enable consistent execution.
- ■ *Clarity for overall operational understanding.* A common language to describe operations and how the interactions occur between different divisions, departments, and other segmentations of an enterprise.
- ■ *Mechanisms for adjusting and driving strategy.* The blueprint to clarify how a competitive strategy is actualized. Processes are the embodiment of a competitive market strategy.

Inside each and every enterprise, processes fulfill these roles regardless of the extent they are understood or managed. In many cases, one or more of the specific process roles are ignored or neglected, inhibiting current performance and the development of future capabilities. When intelligently designed, managed, and executed, processes present a pathway to prosperity—delivering organizational clarity and exacting alignment with an enterprise's mission.

TYPES OF PROCESSES

The universe of processes spans far and wide. Some processes are performed daily (e.g., sales, manufacturing, and distribution processes), whereas others are executed on an annual basis (e.g., year-end financial closeouts or strategic planning). They vary in complexity from simple processes (e.g., distribution of a report) to extremely complicated processes spanning departments and organizational boundaries (e.g., new-product development for a military aircraft). And the value they deliver spans the spectrum from value destroyers to revolutionary game changers.

Inside the walls of all enterprise are groups of processes that share similar labels across the business world, although they are rarely identical matches in their design or execution. The similarities across businesses stem from the sharing of knowledge and tools across organizational boundaries precipitated by employee movement from organization to organization, the commonality of business vernacular in academic and business presentations, widely used software packages, and governmental agency terminology. Some examples of common processes include customer acquisition, credit processing, order acquisition, order fulfillment, employee training, and many others.

Consistent with Adam Smith's theory on the specialization of labor, processes performed by specialized workers are often lumped together in departments. For example, processes focused on attracting, hiring, training, and displacing employees fall into a human-resources bucket. Likewise, processes to manage working capital, raise investment funds, and allocate capital are found in a finance function. These functional groupings map to the conventional organizational chart. In the past decade, reengineering and other improvement methodologies have focused on managing end-to-end processes, and this, in turn, has led to the grouping of interconnected processes into end-to-end processes. *Order to cash*, *hire to retire*, and *concept to design* are examples of end-to-end processes.

Although these labels for processes are useful for identifying common processes, another categorization speaks to the value a process delivers. The processes that create the outputs for end consumers are undoubtedly the most important in any enterprise. As a testament to their importance, this collective group of processes is commonly referred to as the *core value chain*. Each process or activity linked together in the core value chain delivers a part of a finished product or service. But the core value chain does not operate in isolation. Like any living organism, it metaphorically requires food, water, security, and shelter to survive. In a business environment, food, water, safety, and shelter equate to the capital, human resources, facilities, machinery, management, information technology (IT), and all other supporting contributors enabling the core value chain to operate. In a continually operating enterprise, these enabling processes are nearly as important as the core value chain itself. This delineation of processes provides a meaningful segmentation into two categories—primary processes and secondary processes.

- *Primary processes* are the activities/processes that constitute the core value chain. A typical value chain begins with the receipt of raw materials/inputs and ends with the delivery of a product/service to a consumer. As with many overused business terms, the value chain is defined inconsistently. To simplify, the primary processes are where the hands-on work is completed to build the end product/service. Today, value-chain processes receive the lion's share of attention and are designed, documented, managed, evaluated, and improved at a much greater frequency than other processes.
- *Secondary processes* include all the processes supporting the value chain. Although these processes rarely contribute directly to the production of an end product/service, they are integral to the ongoing execution of the value chain. Without them, people are not available to perform the work, there is no money to pay for raw materials, and there are no facilities to house manufacturing. But secondary processes do more than just simply feed the production engine, they include the leadership structures to monitor operations and make strategic adjustments. Based on their unique contribution, secondary processes can be further segmented into support processes and governing processes.
  - ▲ *Support processes* include all the work effort essential to keeping the value chain functioning. Support processes

make the value chain repeatable, scalable, and adaptable. They include processes to create customer demand, finance operations, source raw materials, hire and retain employees, and provide all the necessities of an ongoing concern. They provide the fuel for the core value chain, including capital, human resources, tools, equipment, facilities, and raw materials. From an organizational perspective, supporting processes are usually located in functional departments such as human resources, finance, marketing, and IT.

▲ *Governing processes* provide the directional rules and oversight to manage not only the core value chain but also the supporting processes. Governing processes include strategic planning, workforce management, supervision, engineering standards, quality assurance, audit, risk management, legal, and program/project management. These types of processes provide the guiderails for an enterprise, managing the daily operations within the confines of a predefined structure.

As I will explore later, the type/role of a process is a prime determinant in how it fits into an overall process-management approach. But before we cross that bridge, it is important to understand the life cycle of processes—how they are created, how they change over time, how they eventually become obsolete, and what can be done to make them work better.

PROCESS ORIGINATION AND LIFE CYCLE

By definition, an enterprise consists of the full gamut of processes that allow it to operate as a viable concern. In even the simplest enterprise, processes will number in the hundreds; in major enterprises, thousands.

For a moment, let us reexamine my definition of a process—processes are activities that use inputs to create outputs. This definition does not imply in any manner that most processes are meticulously designed or flawlessly executed. In fact, quite often the opposite is true. Of the multitude of processes in any enterprise, the vast majority are never methodically designed to specific requirements. I call these processes *heritage processes*.

Heritage Processes

Heritage processes are the squatters of the process world. With an unknown origin, they have seemingly been in operation since time began. If there was a plan for their design, it predates institutional memory, and the rationale for the design is long forgotten. Heritage processes are exemplified by the tale of the family Thanksgiving turkey recipe in which the grandmother's directions include the curious step of cutting off the front and rear of the turkey prior to inserting it in the oven. Years later, when questioned about her rationale, the grandmother responds that the step was necessary to fit the turkey in her small oven. For everyone else following the recipe since that time, the step was pure waste, yet executed fastidiously over the years. Heritage processes persist for exactly the same reason—no one bothers to question the rationale behind their design. However being unplanned does not change their importance, heritage processes may well be critical to an output's delivery. On rare occasions, the core value chain may itself be a heritage process. So why do heritage processes continue to exist?

Most employees are well intentioned and perform their duties in a manner that is consistent with their understanding of the job and that minimizes their personal discomfort. In some instances (especially in manufacturing roles), employees receive specific training on the processes and procedures they are to follow. With the passage of time, their performance is heavily influenced by reinforcement mechanisms, including supervisory feedback and observations of their coworkers. Other times, there is a gap when specific instruction is not given. Lacking awareness of how their role fits into the grander scheme, workers mimic their coworkers or invent their own simplified ad hoc processes. And unfortunately, a good number of processes fall into the gap where they are invented by a worker or manager and are accepted as the de facto standard. Lacking formal design, these processes may operate in conflict with leadership's intentions.

Once in place, the heritage processes take root. Although leaders often expect workers to address any deficiencies and investigate opportunities in their work processes, this is rarely what happens. Most workers operate with a limited perspective of how their work fits into the larger picture. They lack the skills, authority, and in many cases the motivation to step outside the confines of their daily role and take on additional work and responsibility. Short of an outright leadership demand, workers continue doing what they have always done—and that includes executing heritage processes. As one might guess, an abundance of heritage processes indicates that strategic adjustments and operational-improvement activities are not being undertaken. And many contemporary enterprises are deluged with heritage processes.

Planned Design Processes

On the other end of the process continuum are the processes designed specifically to meet a business requirement. Someone meticulously planned the performer's steps, the specific inputs, and the attributes of the output. Intent pervades the design—even in instances when it is misdirected or erroneous. Successfully designed processes are the result of experienced process creators who aligned all the aspects of a process to fulfill a specific goal. Once the process is in operation, routines and supporting processes monitor and control the process, vastly escalating the probability that the process will consistently deliver the intended results. ·

Just Poorly Designed Processes

Unfortunately, artfully designed processes are not the prevailing construct today. Based on my observations, the responsibility for process creation or adjustment is often delegated to workers and their managers. In most instances, these individuals lack process training, and their view of the full end-to-end process is limited at best and nonexistent in most cases. Individuals tend to focus on the segment of the end-to-end process that exists in their sphere of responsibility, and do so with minimal interaction and coordination with others. As might be expected, the process fails to deliver as intended or does so inefficiently. Eventually, deficiencies in critical processes become so apparent that they demand the attention of leadership.

Once a process shortcoming is identified, leaders take one of several paths. If the defect is perceived to be the result of poor execution, the process performers may be replaced. On occasion, this solution holds water, and performance improves. An equally likely solution is to identify automation as the solution to poor performance and press forward with a costly and ill-advised IT solution. If you automate junk, you still have junk. It is a well-chronicled fact that the overwhelming majority of IT initiatives today fail to produce an iota of value over cost.

These corrective approaches usually miss the mark because the underlying cause is not execution. William Edwards Deming was a professor, statistician, author, and consultant who won acclaim for his work in developing innovative processes. During his early career, Deming stated that process performance was 80 percent dependent on the design of the process and 20 percent due to execution of the process. At the twilight of his career, he apologetically retracted this statement. He spoke of the grievous untruth he perpetuated. In fact, the design of work was 96 percent accountable for failures, and the slice attributable to execution was only 4 percent.

Despite the "everyone can design processes" belief that is widespread these days, there is an art to process design. It is a skill born of hands-on experience with processes across functions, companies, and industries. Unfortunately, process-improvement skills and experiences are rarely recognized or appreciated. Every day in enterprises, individuals are asked to launch new functions and business lines, yet they lack the requisite process design skills. Not surprisingly, the results fail to meet expectations. It is flawed logic to ask inexperienced workers to design good processes. Experienced process designers are needed. But getting the appropriate process designer is only a part of the battle.

There are a host of reasons why processes are poorly designed— bad requirements, inadequate resources, and an ambiguous scope, to name a few. Additionally, a common culprit worthy of mention is the availability of information. At the time a process is created, details may be lacking in terms of inputs, outputs, or even what exactly the customer wants. In the absence of reliable information, guesses are made. If lady luck smiles and the guesses hit the mark, a good design is possible. More often than not, though, the process is inadequate to the task. Ambiguity is an enemy of process design. A good process design approach accounts for the unavailability of pertinent information.

Even when expert process developers are engaged and the process is fastidiously designed to meet all known requirements, the process still may perform suboptimally. An increasingly common culprit is the influence of external forces on the process's design. *Power-play processes* and *technology-fit processes* are the result of such circumstances.

Processes created in the shadow of political pressure are powerplay processes. During their design, a leader or group becomes aware of the ongoing design effort and engages the designers to voice his or her opinion. Their intentions are not overtly nefarious, but they push to influence the design of the process—most frequently to protect individuals on their team or functions residing in their sphere of control. In many cases, the adjustments they promote are not supported by the business case. If they win, the result is a design encumbered with unnecessary steps, additional costs, or performed by the wrong individual. Examples include additional approval checkpoints to provide for additional monitoring of the process's execution or the handoff of work to another department. But perhaps most discouraging, power-play processes are rarely reexamined after their launch because of the political tug-of-war played during their formation and a general reluctance to rehash the design.

Equally disadvantageous is the technology-fit process. Technology-fit processes result from jamming a business process into a rigid technology solution. Today's leaders tend to equate efficiency with the extent of a process's automation (and then ironically call out IT as the scapegoat after the automated process fails to meet their lofty goals). In the name of expediency, the process is squeezed into the software with minimal or no customization. In many cases, the resulting process can only be executed with manual workarounds. Large-scale technology implementations including enterprise resource planning (ERP) systems propagate technology-fit processes. And in a similar fashion, IT groups are often asked to customize software to exact process requirements. This customization is not only expensive and time-consuming, but it also essentially locks the process into the existing design and destroys any ability for adjustment in the future. Although promises are made to address such shortcomings, cost and competing priorities push the IT department's attention elsewhere. Thus, like the power-play process, technology-fit processes frequently suffer from a lack of ongoing review and adjustment. In hypercompetitive markets, poorly designed processes create a strategic vulnerability because flexibility and adaptability are compromised.

Regardless of how a process comes into existence or is adjusted over time, its intent is to deliver a benefit or fulfill a business need. The consistency and degree to which the benefit is delivered are indicative of a process's performance.

Process Performance

How well a process performs depends on the perspective of the stakeholder. In short, different folks want different things. In the for-profit world, the customer's perspective supersedes that of all other stakeholders. Without sales, a company fails. We know that a customer's purchasing decision is driven by his or her perception that the product or service meets or exceeds his or her expectations. While shopping, the customer considers many aspects of the product/service before making a purchasing decision, including

- *Right product or service.* The product/service is offered with attributes and functions that appeal to the consumer's wants and desires.
- *Right time.* The product/service is available when the consumer wants it.
- *Right place.* The product/service is available at a location that is convenient for the customer.
- *Right price.* The product/service is priced at a value at which the consumer is willing to forego alternative uses of his or her money.
- *Ease of interaction.* The product/service is offered in a manner that is pleasurable or with a minimum of discomfort to the consumer (including time to delivery, customer service, product replacement, tendering options, and servicing).

Sales and loyalty grow when the product/service consistently meets or exceeds the customer's expectations. Hence customer satisfaction is a preeminent aim of the for-profit enterprise. Although the customer is the kingmaker, additional considerations factor in when designing a process to produce a product/service. Aside from satisfying the customer, the next priority is the generation of a financial gain (or profit) from the sale. Whereas the customer's willingness to buy is paramount to capturing sales, managing production costs—both tangible and intangible—and generating a profit are imperative to an enterprise's continuance. Successful enterprises accomplish both—satisfying the customer and simultaneously profiting from this relationship. Although sales and profit get lumped together, they are discrete focuses for a leadership team. Sales are generated when customers agree to pay for a product or service. Profits result when the cost of providing this sale is less than what the customer paid. Together, metrics on sales and profit headline discussions of enterprise performance—and, correspondingly, process performance.

This brings us to the next delineation of process performance. Without a doubt, satisfying customers is job one for an enterprise, but other performance factors are also important—especially when gauging an enterprise's ability to deliver financial returns. The following is a list of the process performance factors that are important to all enterprises. All these factors (and more) are reflections of the design and operation of its processes.

- *Cost.* The full expense to produce, sell, and distribute a product or service.
- *Flexibility.* The capability of a process to adjust to the consumer's preferences.
- *Scalability.* The ability of a process to adjust the volume of outputs to match consumer demand.
- *Compliance.* The conformity of a process to environmental and regulatory requirements as well as societal expectations.

- *Sustainability.* The amount and availability of resources, such as raw materials, used in producing and disposing of process outputs. Sustainability costs and impacts include the full life cycle of the product/service from resource harvesting to the eventual disposal of the obsolete or exhausted product.
- *Safety/risk.* The potential that execution of a process will result in harm to an individual or property. Safety prevention and issue resolution both drive costs.

Collectively, the list of customer-valued attributes mentioned previously and the preceding list determine whether a process is operating effectively and efficiently. Measuring process performance, whether via qualitative or quantitative metrics, provides visibility to issues and uncovers opportunities to adjust existing processes or design new processes to satisfy the customer and to drive financial results. Once target performance metrics are established, someone can be appointed to improve the process. Process improvement is both an art and a science—and a field that has grown and changed considerably over the past two decades.

PROCESS IMPROVEMENT

Processes exist whether they are recognized, managed, or paid the scantest iota of attention by the powers that be. Many managers recognize their importance and develop guidelines for their execution. In this way, they set the stage for consistent execution and monitoring to identify when the train is sliding off the tracks. Such guidelines take many names, including *procedures, standard practices,* or *operational guidelines.* Although guidelines can be effective in monitoring a process to ensure that it operates as intended, the output of a process depends more on the actual design of the process.

This leads to an important question: Who has the content knowledge, experience, and abilities to design processes that are appropriate to the business needs of an enterprise? In most environments, there are a limited number of individuals equipped to the task. In order to address known deficiencies and competitive opportunities, individuals who are unprepared for the assignment are asked to build or improve parts of the organization. As might be expected, their designs often fail to be paragons of efficiency.

Still other enterprises have opted to institute more universal approaches to process management instead of leaving it to individual managers. One retailer sanctioned a super process group to identify process opportunities and assist managers in redesigning their work streams not only to deliver higher caliber outputs but also to operate more efficiently. A sizable number of manufacturing companies adopted the Six Sigma methodology to attack the proliferation of defects in their production processes. One massive financial institution delivered process coursework to its line-level managers and set them loose to drive efficiency across back-office operations. But even in these examples where a large-scale approach to process management was employed, a sizable number of processes continued to reside off the radar screen. These unloved processes include planning, administration, and management processes and other secondary processes often bloated with costs resulting from perpetual neglect. Here lies a major opportunity for cost savings and efficiency gains—and quite possibly the creation of a strategic advantage.

Processes and how they are understood and managed are undoubtedly a significant determinant of success. Processes are the mechanism of value creation. How well the processes are designed and executed translates directly into the quality, functionality, and cost of the finished product. Enterprises that consistently achieve success in their marketplaces do so because their processes are honed to repeatedly produce goods and services of a higher quality and/

or a lower cost than the competition. Sustained performance and competitive advantage are directly correlated with the facility of an enterprise to manage processes across organizational boundaries and through the introduction of improvements to continually refine these production processes.

That said, all processes are not equal in importance. Some processes are critical to an enterprise's success, and without them, a market plunge is inevitable. Processes at the other end of the spectrum are thoroughly wasteful—an expenditure of resources without any discernible gain. To complicate things further, a process's importance may change over time. Today's market differentiator may become tomorrow's norm. Whereas processes are birthed to fill a need, over time, they naturally evolve to accommodate new conditions, new purposes, and new performers. In other words, process evolution does not always occur deliberately. Just as often, performers adjust the steps of a process to make their lives easier. Work is personal to its performer, and employees will find shortcuts unless ongoing reinforcement is there to direct them otherwise. But process evolution aside, ensuring alignment between the desired output and what is actually delivered requires an occasional tune-up to the underlying value-creation engine.

When building products and services, the final output is a factor of both the input(s) and the transformation process(es) followed by the performer. To adjust the final output, there are but two courses of action: (1) change the inputs to the process, and (2) change the process itself. Both approaches have the potential to substantial alter the final output. The choice of which approach to use depends on the scale of the desired change.

The first option is easy to comprehend—acquire different inputs. Switching suppliers and adjusting the specifications of the inputs with a supplier are alternative courses of action. Although substituting inputs may affect the quality, features, or cost of the item, this option is limited in its ability to truly transform an output into something entirely new or with substantially different features. Imagine a shirt. Whereas new fabrics may be used to improve the quality, the shirt is still a shirt. Although input substitution is still a viable and valuable approach, it depends on external business partners and therefore is limited in its ability to be enacted immediately.

The second option—changing the process—brings substantially greater change potential because it harnesses the capabilities of the entire enterprise, including marketing, engineering, and research and development. Returning to the shirt example, the production process might be reengineered to produce other clothing components such as coats or pants. By adjusting existing processes or creating new ones, an enterprise can redesign, reconfigure, reengineer, or rebrand offerings and create an entirely new line of products/services. Processes are the fundamental building blocks of value creation. For this reason, process improvement is the transformative power of an enterprise. It follows that the ability of an enterprise to innovate requires a deep understanding of processes and their capabilities. Although few leaders today think about processes when discussing strategy, all the adjustments in products and services occur at a process level. The challenge is to identify which processes to tinker with and the appropriate adjustments to make.

PROCESS-IMPROVEMENT METHODOLOGIES

As enterprises sought to boost their fortunes over the past two decades, processes became a major focus of improvement efforts. A number of process methodologies appeared on the scene and were adopted by academia, consulting groups, business pundits, and innovative leaders. On a somewhat regular basis, new methodologies arrived on the scene and became the darling of the day while others faded into the background. The most prominent process methodologies over the past decades include such heralded toolsets as Total Quality Management, Process Reengineering, Six Sigma, and Lean. Business literature repeatedly noted that companies using these methodologies were able to reap sizable gains—frequently in the range of 20 to 30 percent increases in productivity. But even when achieving these results, many companies were unable to translate these efficiency gains into increased market share. As a result, the methodologies came under fire and were criticized as being too narrowly focused, too hard to implement, and the wrong medicine at the wrong time. To the contrary, I believe that the fault does not lie solely with the methodologies. Far from it. Many enterprises would be far worse off if they had never launched some form of process-improvement program. What they missed was the full potential of a process focus. These methodologies target the efficiency of processes and neglect to ask the fundamental question of whether a process is delivering the right output. Good process management encompasses far more than examining processes to capture efficiency improvements. It also includes recalibrating process outputs to the all-important customer.

Some of these programs did produce measurable results, including a few that fundamentally altered the norms of an industry. For example, Mutual Benefit Life used Reengineering to make improvements in processing time that fundamentally altered the insurance industry. However, the anticipated financial benefits failed to be realized, and Mutual Benefit Life eventually went bankrupt. There are warnings to be heeded when using any methodology. Misapplying a tool may lead to disastrous outcomes. Solutions can be misapplied or overbuilt. The goal is to avoid these missteps through the correct application of process-improvement tools.

First, methodologies are never intended to replace the thinking and tinkering that needs to occur throughout the analysis, design, testing, and development of a solution. Many teams operate under the fallacious believe that the approach itself will lead to breakthrough ideas if team members just follow the directions assiduously. Based on this perspective, the methodology and its associated templates are a series of checkboxes leading to that pot of gold. But I will emphatically state that the use of any approach as the singular recipe for innovation is courting failure. Breakthrough innovations are the result of aggressive mental investigations from the beginning of opportunity analysis to the testing of the solution. Methodologies are intended to manage the innovation process—*not* be the innovation process itself.

A second mistake occurs when a methodology is used inappropriately or when an alternative approach fits the situation better. These missteps occur because of a general lack of understanding of the specific methodologies and what each does. When a need is identified, managers select the methodology that is familiar to them—which is not always the correct approach. And sometimes the choice is predetermined through enterprise dictates. A number of enterprises have adopted a single methodology for all their improvement projects; for example, Bank of America and Motorola are Six Sigma shops. The obvious risk is using a tool inappropriately. Always use a screwdriver to fasten a screw and a hammer to pound a nail. Otherwise, the results are almost guaranteed to be suboptimal.

The third challenge—and arguably the most dangerous—is the tendency to use a tool or methodology in a limited area of the enterprise, such as Reengineering a part of a finance department. When this approach is undertaken, the finance processes may well become very efficient. However, many of the costs and other shortfalls simply may be shuffled to other departments. For example, one of my clients opted to push the responsibility for the entry of expense reimbursements out of a finance department and to field personnel. In this new model, customer-facing employees out in the field entered their expenses into a tracking tool and then e-mailed copies of the receipts to a payables team. Driven by the finance department, this change did result in lower costs in the finance group because several $20-an-hour payable clerks were laid off. Although the finance payables budget item was reduced, the responsibility simply was shifted to field employees who were compensated at a rate three times higher than a payables clerk. The net effect was a significant increase in overall expense-processing costs, and that was before any impact on sales and customer service was factored in. To mitigate such situations, the design and testing of every cross-functional improvement needs to occur at a system level—accounting for the impact on other parts of the enterprise.

PROCESSES AS COMPONENTS OF A SYSTEM

As stated earlier, every enterprise is comprised of a collection of overlapping, interlocking, interdependent processes. The complete set of processes in an enterprise makes up what I call a *process system* (also called *process structure* or *process network*). Inside this system, work products are built and routed through the network to other teams for further processing and then to customers once the product is finished. There is a ripple effect in such a system. A small change in one location may create waves in other areas—sometimes requiring substantial adjustments to counteract any negative impacts. As a general rule, changes to individual components of the system should always be analyzed, designed, and built with consideration of the full impact on the larger end-to-end process. By using the "big picture" perspective to plan changes, the process designers can mitigate issues by accounting for all known interdependencies. When operating at the process-system level, consideration should be given to these basic tenets of process system adjustment.

- Processes are never singular in existence—they are always a part of an overall system of interconnected and interdependent processes.
- An integrated, holistic view of the process system is necessary to make strategy adjustments or plan efficiency improvements.
- Any initiative begins with a determination and analysis of the processes to adjust and those affected by the improvement to ensure an overall net benefit.
- Metrics to understand the organization's performance must be at the system level and ideally should be focused on the end customer and not internal customers.
- Continual performance adjustments require the management of end-to-end processes and the continual alignment of tangential subsystems and supporting processes.
- Any work on processes must consider and include process interdependencies. Outputs of processes that initially appear to be waste may serve as inputs to downstream processes.
- Changes to cross-functional processes require communication and coordination with affected teams to avoid any negative impacts. This is especially true when deploying a new solution that affects other parts of the enterprise.
- An efficient mechanism to allocate resources across the entire process system is required—or waste will appear in the form of wait times, excess capacity, inventory obsolescence, and underutilized resources.
- Systems work requires iterative testing and analysis because chaos exists in every process system and makes exact forecasting of outcomes challenging, if not impossible.

Improving a system is significantly more complex and difficult than improving an isolated process. The complexity of the system obscures causal relationships, making forecasts of outcomes inaccurate and misleading. Unfortunately, this discourages many leaders and managers from working at the process-system level. However, moving forward with adjustments without a link between action and reaction makes strategy, innovation, and any improvement effort pure guesswork, and only with the greatest luck will the results align with intent. In a related manner, the relative simplicity of an isolated process lures managers to take the easy road and ignore the larger system. In some instances, the unintended consequences will be significant and cause a large amount of rework. And unfortunately, once burned, leaders acquire an aversion for "risky" innovation efforts—eroding their support for future process endeavors.

The right approach is to understand and embrace the process system. Unintended consequences still may occur, but the risk is manageable. When working on a process, always conduct an end-to-end assessment of the impacts of the improvement. Test the solution in a laboratory environment, and run it through multiple iterations under different test conditions. Pilot the solution, and examine closely how other processes and functions are affected. With these actions, the end result and most of its consequences can be anticipated—allowing for any uncovered detrimental outcomes to be addressed prior to a larger deployment.

Despite a history of launching multitudes of improvement efforts year after year, many enterprises ride a bumpy road when adjusting their offerings and building platforms for future growth. The difficulty stems from many areas: a lack of customer focus throughout the organization, a deluge of reports and metrics that misrepresent the true situation, structures and processes that are misaligned with the strategy and are constantly under repair, and fires raging on a regular basis, stealing leadership's attention and preventing leaders from methodically planning innovation efforts. In this environment, the big issues are pushed to middle managers for resolution—leading to the subsequent launch of teams and initiatives. But are the efforts directed at the critical problems or just the immediate ones? Does the enterprise really understand its operations? Does the enterprise even truly have a market strategy, or is it just copying the competition?

In a nutshell, the real problem is that until now, there were no structures or methodologies to address the full scope of challenges confronting leaders. And this is what is needed—a holistic framework on which not only to manage an enterprise but also to react appropriately to changes in customers, competitors, and market conditions. Fortunately, there is now an approach that uses the construct of process to identify, define, and systematically manage adjustments to an enterprise's operations—leveraging toolsets and methodologies already well known to many and pulling them into a cohesive system for running and innovating an enterprise, what I call a *process-based approach*.

4

Innovation via a Process-Based Approach

I n prior chapters we examined the historical challenge of keeping pace in an ever-changing world and learned that innovation is the key to the success and indeed the continuance of any enterprise. We studied the four facets of an innovative enterprise—customer focus, strategic planning, operational improvement, and initiative management—and how they are integral to overcoming market and customer shifts. We examined how making fundamental strategic or operational adjustments requires analyzing and reinventing work activities at the process level. In other words, intentionally adjusting a product or service requires altering the routine steps performed by workers. And so that the adjustments are not blind shots in the dark and account for possible repercussions, making a process adjustment requires a thorough understanding of the causal relationship between the end product and both the process inputs and the process used to manufacture the outputs. By leveraging these relationships, it is possible to forge a link between the intended outcome and the production mechanism of an enterprise. Unfortunately, in larger enterprises, the causal relationships are increasingly more complicated to predict because of the growth in the number of processes and their interconnectivity. At this point, we have a good grasp on what needs to occur to improve an enterprise. The question now is how to build an innovation machine—one that refreshes itself as conditions unfold and drives an enterprise in the intended direction.

When viewed in isolation, processes are artificial constructs. Visually depicted as a collection of lines, boxes, and words in training manuals, guidebooks, and other materials, they are by themselves lifeless and valueless. People—employees, associates, partners, volunteers, and others—are the fuel that makes things go. People design, build, implement, and perform the processes. Although we talk about process as the foundation of work, it is the employees who perform the processes and make things happen.

At an elementary level, innovation involves devising new ways to organize and direct workers to deliver outputs with unique attributes. Although it comes as no surprise to anyone in corporate America, the prevailing hierarchical organizational structure does a poor job of aligning people with value-added work. It is a structure to control and supervise workers, but it has limited connectivity to the actual work. In addition, it fails to identify oversight for the large numbers of cross-functional and end-to-end processes operating in most enterprises and thereby contributes to the silos and knowledge-gap deficiencies we identified previously.

In a process-based approach, work activity is defined in terms of processes. Additionally, change is defined in terms of adjustments to a process or set of processes. It follows, then, that the most logical way to organize employees is based on their relationship to a process. *Process management* is the practice of organizing a workforce around the processes in an enterprise.

The Concept of Process Management

At the epicenter of process management is the role of the *process owner*. In contrast to managers presiding over a functional area today, process owners are responsible for all facets of an end-to-end process. This includes the daily execution of the process, oversight and direction of its performers, implementation of strategies that affect the process, and maintenance of the ongoing efficiency of the process. Because a good number of the most important end-to-end processes span departments and other organizational boundaries, a process-management structure uniquely assigns individuals to the work activities most critical to an enterprise. To further clarify this important role, the following is a list of responsibilities typically associated with process owners:

- Manage the execution of an end-to-end process on a daily basis.
- Adjust processes to support strategic and operational improvement initiatives in collaboration with other process owners.
- Continually seek methods to make processes more efficient without harming—or in concert with—business partners.
- Act as representatives of the process and be able to speak to all facets of the process—resources, costs, ongoing and future improvements, metrics, and so on.
- Train and mentor process performers.
- Understand and speak to the resource requirements of the owned processes.

Process ownership vastly expands the ownership of innovation from a select few individuals (i.e., a management committee) to a wide swath of individuals spanning every part of an enterprise. In this model, the process owners oversee the adjustments to their process—and this brings implications to the ownership and execution of improvement initiatives. I will cover the ownership of initiatives at a later point. Regardless of ownership, initiatives managed in this environment are created and managed using a process-based approach.

THE PROCESS-BASED APPROACH

Today's enterprises struggle to make fundamental changes to the way they do business. When the market is stable, the path forward is easy—with only minor obstacles to skip over. But when big change arrives (i.e., think of the disruption to book sellers when Amazon started selling books online), these same enterprises flounder and bleed market share. In many cases they will be lucky to survive. They struggle to right the ship, even when they know exactly where they need to go. Most companies today rely on some variant of a strategic planning process to identify and schedule improvements. Typically, the senior leadership team hatches a number of well-intentioned directives with poorly defined scopes, wild guesses as to their predicted benefit, and nebulous responsibility for their execution. From the inception, problems arise. For a moment, let us ignore the fact that some initiatives are strategically misguided—just plain aimed at the wrong target. Assume that we are looking at an initiative that is perfectly dialed in strategically. Taking it from that point, the initiative receives the leadership's blessing and is pushed to the dock for launch. And the problems start. Because many initiatives are birthed behind closed doors under a veil of corporate secrecy, the first challenge for any initiative owner is to get a firm grasp of the true intent of the initiative. What did the leaders/strategists really intend to accomplish? Common sense would suggest that the process owner should circle back to the initiative creators for additional detail. In today's corporate environment, though, requesting clarity is equivalent to admitting you are lost—and not the right person for the job. And quite frankly, even when clarification is requested, many senior leaders are at a loss to give guidance because initiatives are frequently hatched in a vague high-level form. As a result, initiative owners are left to push forward with meager direction. Only the superelite change agent has even a ghost of a chance of making such an initiative successful. It is hard to reach a destination when you do not know what it is. This brings us to a key concept and a huge benefit of a process-based approach—*clarity of intent.*

In lieu of guesswork, in a process-based approach, initiatives of all types—strategic and operational improvement—are translated into the language of process. In other words, every initiative specifies the processes involved and the projected adjustment to each individual process. To illustrate this way of defining improvements, consider a sales initiative to expand the distribution of a product through the use of a network of independent distributors. To start, we break down the initiative into its process components. Such an initiative would definitely include the creation of new sales processes aimed at distributors. It also probably would require the inclusion of the new distributors' delivery spots in transportation and distribution processes (i.e., route management, inventory management, and so on). Additionally, sales management, pricing, customer service, and other processes would need to be involved in the execution of the initiative. By breaking down an initiative into process adjustments, the ambiguity and confusion around the intent and composition of the initiative are immediately dispelled.

But this approach comes with a formidable change to corporate cultures. Identifying the processes to be adjusted and the expected results requires initiative planners to get dirty—to get elbow deep in the specifics. And it's about time! From my perspective, this level of detail is a necessity for innovation planning. Strategists and senior leaders cannot continue to sequester themselves away from the details of how their enterprises operate yet still keep their hands on the knobs and levers that propel change. The responsibility to lead an enterprise necessitates leaders driving the innovation process. They simply cannot be effective operating at 25,000 feet when real change takes place in the trenches on the front lines where the goods and services are produced and sold to customers. This approach yields several immediate benefits.

Translating Change Efforts into Process Terms Provides Several Benefits

- The scope of an initiative is clearly identified leaving no room for confusion and eliminating the need for initiative owners to beat the bushes for clarity.
- The very processes (and therefore process owners) that must be included in the design and execution of the initiative are spelled out in the initiative itself.
- Using processes as the basis of communicating change initiatives creates a foundation for initiative management. This is a topic deserving of a significant amount of attention and will be covered in detail in Chapter 7.

THE PROCESS-FOCUSED ENTERPRISE

A *process-focused enterprise* is an organization that has adopted the process-management philosophy to manage its resources and also defines improvement initiatives in the language of process adjustments. Operating as a process-focused enterprise requires the implementation of three key elements—a process structure, a governance structure, and an innovation plan.

Returning to the story of the Digital Equipment Corporation (DEC) in the Preface—where senior leaders lacked a basic understanding of their company's operations—this case exemplifies a widespread problem. Any effort to improve the operations of an enterprise requires the change agents to fully understand the area to be adjusted—the process used, the performers, the cultural influences, the metrics employed, and so on. In the absence of such detail, the improvement effort is a blind shot with its success largely dependent on luck. And although the "Ready, Fire, Aim" approach is alive and well in corporate America, improvement efforts based solely on blind shots tend to miss the mark. In all likelihood, the effort will be directed at the wrong target or its aims will be impractical based on the current capabilities of the enterprise. Expecting leaders to possess this ground-level awareness on their own invites the risks of personal interpretations and variations in the completeness of their knowledge. If planning is to be productive and grounded in reality, an accurate map of the enterprise's operations from a ground-level everyday view is essential. To provide this perspective, I recommend the creation of a process-based view of the enterprise—what I call an *enterprise process blueprint*.

An enterprise process blueprint depicts all the major processes that together give an enterprise the ability to operate on a continual basis. It is analogous to a traditional organization chart in its use as a bird's-eye view of an organization, but there are significant differentiators. The first glaring difference is inclusion of the customer. The customer is the reason for any enterprise's existence and the arbiter of its success. Why would the customer ever be excluded from any presentation of an enterprise's operations? And an enterprise process blueprint includes all the major processes of the enterprise. It is a picture of the work that occurs in an enterprise, not the command and control lines used to organize people. In this way, it more accurately represents how work products are created. Depending on the intended use, subprocesses (i.e., minor processes supporting a major process) may be noted beside the major processes to yield a more comprehensive operational view. Arrows are used to identify the flow of work throughout the organization, as well as the customer connection points. Supporting processes can be included, although they may be placed in an area separate from the core value chain to reflect their supportive nature. When external partners and suppliers are critical to the creation of value, they also need to be included on the blueprint. An enterprise process blueprint is the highest-level view of the *process structure*—the full complement of processes that collectively allow an enterprise to operate on a continual basis. As such, the enterprise process blueprint is the starting point to identify any process adjustments required to implement a strategy or operational improvement.

Building an enterprise process blueprint is by itself an illuminating experience for leaders—especially creating one for the first time. It simplifies operational components into an easy-to-understand picture and offers a clarity of purpose previously lacking in most enterprises. (Chapter 5 covers enterprise process blueprints and their creation in detail.) Once complete, enterprise process blueprints become an indispensible organizational view—framing debate and discussion for operational and strategic planning. But the blueprint is only a piece of the puzzle—there is more to becoming a process-focused enterprise than simply knowing how things operate today.

Mapping the process structure to the traditional supervisory structure (i.e., how people are organized in the enterprise) creates a conflict. The two structures align very poorly. Remedying this disconnect is just common sense. Use a single structure to identify the flow of work, and use this same structure to organize workers. This is the second piece of a process-focused enterprise. Switching from a boss/employee-based organizational structure to a process-based organizational structure involves a paradigm shift from the prevailing command and control structures to value-creation structures.

This shift aligns workers with value-creation activities and sets up a new leadership model—what I call a *process-governance structure*. In this model, managers are mapped to processes and are called *process owners*. Where individuals were previously ordered into functional departments, they are now assigned to processes. Mapping individuals to processes requires jettisoning some of the outdated human-resources practices founded on the functional department and instituting new approaches to training and managing employees. The overall process-governance structure, including new roles and practices, will be covered in detail in Chapter 6.

The third and final piece of the process-focused enterprise is what I call the *innovation plan*. The process structure is the blueprint of how an enterprise operates. The process-governance structure organizes people to this structure. The last piece, the innovation plan, is the mechanism to manage and coordinate improvements to the operations of the enterprise. It is how enterprises get updated and repositioned to succeed in a changing environment. At any given time, employees are heads down with the day-to-day activity of manufacturing products and performing services. Raw materials are purchased, transformed into consumer-desired products, and sold to customers. In other parts of the enterprise, the information technology (IT) department codes software to automate processes and capture information for leaders, human resources hires new employees and files reports to comply with government regulations, finance creates a budget and obtains investment funds, and the list goes on. This everyday work is central to an enterprise. It cannot be delayed or ignored. However, as the world changes, the enterprise must as well. Here is where the innovation plan fits in. As the enterprise leaders plot a future course for the enterprise, they actualize these plans in initiatives to either build new products/services or adjust existing products/services. Initiative management includes all the activities to manage improvement initiatives throughout their life cycle—from initial identification to eventual execution. As defined previously, an innovation portfolio is the full set of all initiatives (both strategic and operational improvement) in an enterprise.

By actively managing the portfolio of initiatives, leaders can rank and execute the initiatives in a manner that increases the total benefit delivered by the collective set of initiatives. This is accomplished by expediting higher-value initiatives and culling initiatives that are no longer predicted to deliver value. Prioritizing the initiative portfolio in this manner focuses attention and resources on the activities that are most beneficial to the enterprise. Once the prioritization is complete, the end result is an innovation plan.

Because only a miniscule number of enterprises have taken the step of formalizing a methodology to continually prioritize and execute all initiatives, the enterprises embracing this concept reap a major advantage over their competitors. The end goal of this book is to use the process-based approach in alignment with the four facets to build a continuously innovating enterprise.

INTEGRATING THE FOUR FACETS INTO A PROCESS-BASED APPROACH

As identified previously, the four facets of an innovative enterprise are elements that are integral to a growing and evolving enterprise. Depending on market conditions, one or more of the facets may take a position of dominance. But, for enterprises to prosper continuously, each facet must be in existence.

Before broaching the topic of systematizing innovation, let's briefly review the four facets of innovation (Figure 4.1). The first facet is *customer focus*. Understanding customers and their needs, desires, and behavioral patterns is critical to building products and services with market demand. By gathering information from

Figure **4.1** The focus and outputs of the four facets of innovation.

	Customer Focus	Strategic Planning	Operational Improvement	Initiative Management
Focus	• Internal Feedback Loops • External Research & Benchmarking • Trend Analysis • Customer Analytics	• Competitive Assessment • Core Value Chain Analysis (Capabilities Assessment) • Game Theory	• Process Transformation • Technology Development • Organizational Analysis • Structural Analysis	• Initiatives Developed • Overall Initiative Management • Resource Allocation ($, headcount, focus)
Output	• Customer Processes • Customer Perspective	• Strategic Initiatives	• Operational Improvement Initiatives	• Coordinated Initiative Plan • Resource Allocation

myriad sources (both internal and external to the enterprise) and by overlaying this information with general macro and micro trends, customer profiles can be built for a product/service family. Through this exercise, an enterprise can identify the desired attributes of a specific product/service.

The second facet, *strategic planning*, builds on the customer-focus facet. By knowing the customer-desired attributes of a product/service, a planning team can identify promising market opportunities. By evaluating these opportunities in light of anticipated competitor responses, the planners predict the consequences of their strategic actions and commit to those generating the greatest return. The activity to capture these strategic opportunities is bundled into strategic initiatives.

The third facet is *operational improvement*. Whereas strategic planning addresses market opportunities to capture sales, operational improvement is focused on the enterprise's inner workings. It includes activity around developing future capabilities; making the core value chain more scalable, adaptable, and flexible; reducing costs; improving quality; and reducing cycle time. After the opportunities are identified, they are documented as operational-improvement initiatives.

The final facet is *initiative management*. The initiatives born of the strategic planning and operational efficiency facets are consolidated into a list called the *innovation portfolio*. The initiative-management team evaluates and prioritizes each initiative using a methodical approach and with consideration of dependencies and collaboration opportunities. Each initiative is then evaluated and ranked in an order to maximize the collective return of the portfolio. Based on the order, the enterprise allocates resources and attention to the most promising initiatives in the portfolio.

The four facets of an innovative enterprise are the critical elements of an enterprise's innovation cycle. We know that processes are the single best way to align planning with daily execution. It follows, then, that merging the four facets with a process-based approach creates an optimal innovation approach. To understand exactly what this means, it helps to walk through the four facets as they contribute in a process-based approach environment.

Customer Focus

The customer-focus facet includes collecting information on existing and potential customers with the intent of generating insights to build customer loyalty and further grow sales. However before the ship leaves port, the immediate question demanding an answer is, "What customer?" Today's organizational structures range in size from sole proprietorships to multinational conglomerates selling radically different products to a host of customer groups. Particularly for larger, complex enterprises, painting a picture of the customers for a specific product/service offering requires a bit of an investigation.

There already exist droves of books, webinars, articles, and thought pieces on customer segmentation that do a very good job of identifying and segmenting customers based on their purchases and their use of products. For this reason, I will address customer segmentation only in a very cursory manner.

Many enterprises categorize their customers based on a legacy segmentation approach or even mimic how popular surveys and journals segment the industry—for example, car classes such as economy, luxury, and sports. When meaningful categorizations do not exist, the first step is to define groups (what are commonly called *product/service families*). Product/service families are based on any number of factors, including whether the products/services they purchase are complementary, whether they are substitutes for each other, or whether the different products/services are sold to similar customers. Any number of factors might determine a product/service family. Regardless of the basis for delineation, the true purpose of this exercise is to laser in on the set of customers for a product/service family. In today's rapidly changing product markets, customer segments may be fluid. Be flexible in the mapping. As additional information surfaces, it may make sense to combine product/service families, but it never hurts to start with a more detailed and segmented view.

Once the product/service families are determined for an enterprise's offerings, the real work begins. For each product/service family, a detailed view of the customer is created, including the product/service attributes valued by the customer, how the customer shops for the product, how the customer uses the product, and how the customer's preferences are changing. Kano analysis, as presented in Chapter 2, is a good model to analyze customer preferences. Another tool is to use the process performance factors identified in Chapter 3 to identify differentiations. The goal is to build a complete picture of the customer that encompasses all his or her significant preferences, behaviors, motivations, and attitudes. When creating this picture, extend the identification of customer attributes to include any latent or expected future needs. Otherwise, the customer information is pegged to the past. Innovation is calibrating the enterprise toward the future customer—building products, services, and experiences that are enticing to future buyers. As with any analytic endeavor, the biggest stumbling block in this exercise is gathering high-quality information. Again, no single source suffices. The best customer perspectives are pulled from multiple unique sources and consolidated together into a singular customer perspective for each product/service family.

Earlier I mentioned a number of sources that are useful to gather customer information, including cash register analytics, external research providers, internal surveys or focus groups, and internal feedback loops. In a process-based approach, the information sources are largely the same, but the contribution from insights generated through internal processes plays a much larger role. For example, the customer is placed at the apex of the enterprise process blueprint. Not only does such a prominent position loudly communicate the value of the customer, but a thorough blueprint also identifies the connection points where the customer and the enterprise interact. Each of these customer contact points is a potential gold mine of customer insights—if only they are harvested. If there is presently no formal feedback loop from these points to a single customer information team, rectifying this oversight is priority one for a customer-focused enterprise. Pulling customer information from the front lines is always the best way to gather current and insightful data on a customer's actions, motivations, and future intentions.

Despite their importance, a major shortfall for many enterprises is the utter lack of understanding and management of the processes connected to the customer. Companies that devote time and resources to delivering more value for their customers end up being the strategic superstars of their industry—the disruptors of the status quo.

They delight customers and nab market share while the competition's attention wanders.

This brings us to one of the most overlooked but most promising opportunities in the strategic realm—managing the customer's processes. Customer processes include all the ways customers interact with an enterprise and its products. To give a short list, customer processes include how they shop, purchase, use, resupply, repair, and dispose of a product. We know that positive experiences throughout the product life cycle build loyalty and translate into future sales. Remember how most enterprises attempt to innovate—via a strategic planning process. Strategic planning processes are generally internally focused—a list of directives that focus on adjusting market position and operations and building future capabilities. However, this approach to strategic planning ignores the most powerful weapon in the strategy arsenal—the ability to reengineer the customer's processes and thereby create a wholly new (and hopefully superior) customer experience.

The iPod is a great (although overused) example of a product whose company (Apple) understood the customer's processes and reengineered them. Apple knew that customers wanted conveniently available music and that they were more than willing to cut out a store visit if they could buy it online. With this insight, Apple launched iTunes, an online store for music. Now customers are able to search for, sample, and buy music from the comfort of their home or wherever they happen to be. Customers valued this convenience as well as the elimination of the hard product (compact discs). By providing a superior value statement, the music distribution industry was transformed forever.

Apple was able to make such a leap in delivering value because it viewed the customer not as simply a collection of current perspectives and slices of information but rather as a living entity with evolving wants and needs—some very visible, others more abstract.

Putting structure to this undertaking requires a process viewpoint—a detailed awareness of the intricacies of the customer's actions. Only with this level of understanding of the customer can game-changing innovations be designed and brought to life.

To attain this level of insight, someone in the enterprise must not only map and intimately examine the customer processes, but he or she also must be a master communicator and make this knowledge widely available to strategists, planners, and process designers. Innovation is vastly more successful when it is a team effort—not an activity left to a select few isolated in conference rooms about the corporate headquarters.

To hasten the development of customer processes, I find it helpful to segment customer processes into two categories—*shopping processes* and *product/service usage*. Shopping processes include all the steps a customer goes through prior to making a purchase. These steps vary greatly depending on the type of product or service purchased. For example, buying a hose from a local hardware store is significantly different from the process involved in remodeling a kitchen. Although a hose purchase might just be a stop on the way home, a kitchen remodel requires a much larger investment of time and resources. It entails planning the kitchen, choosing materials, soliciting bids from contractors, and a host of other activities. The following is a short list of customer shopping processes:

- Defining or clarifying the want or need
- Gathering information about the product/service and purchasing process
- Understanding the product/service processes and how they can be used, serviced, and so on
- Obtaining pricing for the product/service
- Designing or configuring the product/service for the customer-specific situation
- Receiving the product

Once the purchase is made, the product/service might require further processing before it is ready for use. And with continued usage, additional servicing, repairs to damaged components, or refills are often necessary. All these activities related to the setup, use, and disposal of the product are also customer processes. Several examples of usage processes are as follows:

- Setting up or installing the product
- Getting assistance from the company, written directions, or other sources
- Resupplying, reordering, reloading, refurbishing, and repairing the product
- Disposing of the product

By combining knowledge of the customer's processes with insights gleaned from other sources, a sizable amount of customer detail is available to guide product/service design. Simply injecting rich customer insights into the planning process is a major step forward in comparison with the strategies generated from half-baked guesses that are abundant in strategic plans today.

This brings us to an immediate benefit of a process-based approach. If an enterprise wishes to build unwavering fans in its customer base, a major opportunity is to manage the connection points with customers in the same way internal handoffs between functions are managed. The aim is to connect internal processes with the customer's processes in such a way as to create a delightful customer experience and to do it consistently. This attribute of customer focus is frequently labeled as *easy to do business with* (*ETDBW*).

Today's companies tend to operate in terms of events—the busy season (e.g., around holidays for retailers), the budgeting season, and the strategic planning process. Each of these events/activities has predecessor activities such as the compilation of prior results, market studies, or other pertinent information. In the name of convenience, key planning processes are scheduled to accommodate the arrival of this information. But customers and competitors are continuously changing, so collecting information at a single moment in time ensures that you have a perspective that is meaningful only for that singular moment. There is momentum to change. Sometimes it is accelerating—other times it drifts. The short scoop is this: *collecting customer and competitor information should be an ongoing process.* It does not start or stop with planning cycles. It is ongoing and dynamic. As enterprises expand their capabilities to gather and analyze customer information, teams need to be assembled not only to build the *voice of the customer* but also to represent the customer throughout planning and innovation activities. In the absence of such a role, the cacophony of internal leaders clamoring for their individual needs drowns out the voice of the customer. And when customer knowledge is limited, any strategy is a glorified game of chance. Blind luck occasionally may trump diligent planning and preparation, but the enterprise that is zeroed in on the customer is at least aimed at the right target.

Strategic Planning

The predominant practice in corporate America is to execute a strategic planning cycle on an annual basis. As part of this exercise, customer data is studied, competitive tactics are discussed, and opportunities are investigated—at least theoretically. In many instances, this planning cycle overlaps with a budgeting cycle. As the curtain is pulled back on the prevailing practices, the real truth is that in a solid majority of companies both the budget and strategic plan are created using some variation of the *SALLY method (same as last year plus a little more)*. This is how it often works. Templates are distributed to functional leaders so that they can identify their big opportunities. Some are strategic; many are not. The completed templates are aggregated together into one big plan and blessed by the powers that be.

Such an approach is riddled with flaws—and here lies a major reason why big corporate America is losing ground to smaller and more innovative players. Executing a strategic process on an annual cycle is equivalent to giving a gift to the competition. In effect, an annual cycle indicates that the attention of leadership is focused elsewhere (usually on financial results or investment analysts)—not on the customer or the competition and definitely not on an investigation of market opportunities to grow market share. This leaves an enterprise strategically vulnerable during a good portion of the year—abandoning the customer to more nimble competitors who opt to recalibrate their strategy on a more frequent basis.

An equally significant issue with most strategic planning processes is the diminution of the customer to a relative level of obscurity. Although it sounds insultingly basic, effective strategic planning starts with the customer. Customers should not be an afterthought, written into the picture after initiatives are determined. Customers are the reason for the initiatives. They should be where the improvement ideas are rooted.

A final issue with strategic planning today is the lack of foundational structures and processes for strategic planning. Many leaders today hire top-dollar consultants to develop their strategic plan. This is the path of least resistance—cut a check to receive a slew of ideas based on studies of market information. There is a safety valve for leaders with this approach. If an esteemed firm did the job, the recommendation has to be right, right? However, to date, I have never heard of a leadership team hiring consultants to develop their internal strategic planning process. And that is the hard part—developing a system that continuously reenergizes the enterprise.

Enterprises come in all shapes and sizes. This complicates strategic planning because strategies need to be created at the appropriate level. For holding companies, there may be strategies at the holding-company level, and there may be strategies for each of the distinct businesses that the holding company owns. A diversified company may require several strategic planning teams to develop unique strategies for each business. Likewise, a sole proprietorship may require only one strategy if it is a single-threaded business.

In a process-focused enterprise, the difference is that the strategic planning process receives the same attention as any other major process. It is assigned a process owner and is continuously evaluated for improvement. Instead of reinventing the process every year, leaders can focus on the customer, the competition, the market offerings, and the capabilities of the enterprise.

Assuming that the enterprise builds feedback loops from frontline performers and other internal/external sources, this information is passed to the strategic planning team, whose members then generate ideas on how to best position the enterprise's offerings in the market. When analyzing the process's success, the first question is whether the strategists have sufficient information to execute the strategic planning process. Is data on current and prospective customers rich and insightful? Has due diligence been completed on the competition and how they might react to various strategies? The extent to which information flows throughout an enterprise is directly correlated with its innovation capabilities. To ensure that crisp information is available for the strategic planning team, the customer-research function is frequently placed in the same organizational area as strategic planning.

At this point in the innovation cycle, product/service families are identified, customer segments are associated with those families, and customer-desired attributes for each product/service offering are captured. With a solid base of information, the next step is to begin identifying strategies. Every product or service has a strategy, regardless of whether it was created subconsciously on the whim or meticulously developed by a specialized team. In effect, standing pat is a strategy. Although there are multiple approaches to create strategies, strategy creation is an intuitive process—the formation of ideas based on knowledge of the customer and the prediction of competitor responses. As these strategic ideas surface, the first step is to complete a back-of-an-envelope analysis of the opportunity—a quick calculation of the potential value of the idea. Although simplistic, this estimate verifies that the strategy has potential and prevents the waste of time and effort on unproductive ideas. This is the first gate for strategic creation—is it significant enough to deliver value? With a rough number in hand, the next step is to consider the competitive ramifications. How will the competition react?

There is no sure-fire way to predict competitor responses. My recommended approach is to study the competitors' historic reactions and gather the available information on their leadership, financial state, ownership, and prior competitive strategies. At that point, an educated guess can be made of the potential reactions. The end goal of this exercise is to predict how competitors' responses might change the expected benefits of a strategy. But predicting the competitors' reaction is always challenging.

When assessing competitive options, one useful tactic is to map the competitor processes to understand their capabilities. How hard would it be for competitors to emulate your proposed strategy? Can the competition even respond? Are there other weaknesses in their strategy to exploit? For example, do prices need to be lower to win customers? Are there additional components to the solution not yet ready for market?

Assuming that the decision is to continue exploring the strategy, the intended outcomes need to be further scrutinized to confirm that they are viable given the enterprise's capabilities (i.e., Can we build it?) and to confirm that the consumer will embrace the output (i.e., Will they buy it?). As mentioned previously, value is created through a group of processes called the *core value processes*. These core processes transform raw inputs into products and services that are valued by prospective customers. Success occurs when customers choose to part with their hard-earned money and actually purchase these products/services instead of competitive offerings. For this reason, these value-creation processes are the embodiment of an enterprise's strategy. They represent the unique manner by which an enterprise repeatedly creates products and services for its customers in order to generate financial returns. The adjustment and innovation of the value-creation processes are strategy at its most elementary level.

To clearly define a strategy, it is helpful to identify not only the affected/adjusted processes but also the specific outputs delivered by each process. One straightforward way to depict the flow of outputs from the value chain to the customer is to build a pseudo–process map. To build this view, trace the desired outputs from the customers back to their source processes. In Chapter 9, detailed guidance will be provided on how to associate outputs (and the desired attributes of those outputs) with the processes that deliver them. The outcome of this mapping exercise is a crisp view of the process changes/ additions to be completed to take advantage of the opportunity. It identifies adjustments to the core value chain processes, supporting processes, and customer touch points. This picture eliminates any confusion as to strategic intent and mitigates instances when managers and line-level associates are forced to decipher leadership dictates before they can proceed with the initiative. And beyond the simplicity and clarity gained, putting strategies into this framework builds a shared awareness of a strategy's implications across stakeholder groups. An understanding of the end game is critical to a successful strategic deployment.

Any experienced strategist knows that strategy creation is not an exercise in which ordered steps are executed to arrive at a specific destination. It is an iterative process based on available information and insights that may well lead to a different point than was previously thought. It is creative brainstorming in front of a white board—a free for all where ideas are tossed up and developed and then either discarded or studied further. To confirm their utility, the proposed ideas require verification from customers as well as internal resources and business partners. And then there is more spin, more analysis, and more redesign until the white-board proposal delivers a customer-valued output that can feasibly be produced by the enterprise to capture a market advantage. Defining opportunities in terms of process adjustments creates a specificity and clarity beyond anything available today. As many midlevel managers will attest, the time and energy expended on expeditions to understand executive intent are beyond madness today. Simply using processes as the language of change can transform an enterprise from a strategic laggard to a market-share stealing machine.

At this point in the strategic process, the key stakeholders including process owners, business partners, and select customers have vetted the opportunity and confirmed its validity. The end-state objective is crisply defined, and the requisite processes improvements (i.e., adjustments or new processes) are known. The strategic objective is nearly ready to be handed off to a project team for execution.

The first step is to break the overall work effort into initiatives— or executable divisions of work. Strategic opportunities come in all sizes—some span the process structure, whereas others have an impact on only a select few individuals. However, there is a limit to the size and scope of initiative that can be executed. When initiatives become overly large, they invariably stall and are either scaled back or abandoned altogether (usually after months or even years of wasted time and cost). For example, a major retailer undertook a massive reengineering effort. As time passed, more stakeholders joined the discussion and added their requirements to the overall program—vastly expanding the program's scope beyond the initial intentions. Several years after the inception of the program, the daily spend on program resources approached a million dollars. With all the additions made to the program, the original business case was defunct—lost among all the add-on work. Work efforts continued, but the end state was just an undefined glimmer off in the distance. Eventually, the program ground to a halt after a series of missed deadlines led to a more comprehensive review of the business case. Shortly thereafter, the company underwent a series of layoffs and drastically scaled back the program to something approximating its initial scope. The lesson here is that initiatives can be too big to implement. Enterprises and their workers are unable to handle change of such a grandiose size and scope. Big change requires small steps.

When building initiatives, the guiding principles are twofold—create initiatives that provide meaningful value and that are achievable. Every initiative should be a discrete piece of work that moves the ball a step forward. Where work activities are interconnected, bundle them into a single initiative if size permits or link multiple initiatives together to build towards the desired end state. With the planned work segmented into manageable groupings, the final step before the initiatives are transitioned to a project team is documentation. To accurately define an initiative, a set of basic information is required:

- Initiative name
- Initiative intent/goal
- Initiative benefit
- Expected outcome in terms of the outputs
- Processes affected significantly
- Other processes affected

With just this information, teams can understand the initiative's scope and projected benefit. A deeper perspective on initiative development will be provided in Chapter 9.

One final note on using a process-based approach for strategic planning: strategic planning is a process. Like any other process, it deserves attention and active management—that is, a focus on its results and continual improvement. It requires conscious design, continual analysis to improve its effectiveness and efficiency, goals for improvement, training, knowledge sharing, and all the other "love" that core value chain processes enjoy. Unfortunately, the strategic planning process is rarely the focus of improvement efforts in corporate America. But here is a major opportunity for the innovative enterprise to get a jump on the field. In a very similar fashion, operational improvement provides a mechanism for enterprises to best the competition by paying attention to other areas neglected by many enterprises.

Operational Improvement

Operational improvement rarely receives the attention that strategic planning does. It simply fails to engender the same pizzazz as do efforts aimed at revenue growth or capturing market share. Today, when internal operations do receive attention, it is because a barrier was discovered that prevents leaders from taking a chosen path or tough times have descended and cost cuts (i.e. layoffs) are needed to meet financial goals.

In stark contrast to their reputation, operational initiatives are a powder keg of potential. First and foremost, efficiency is a driver of profitability. As sales are registered, well-managed processes deliver outputs at lesser cost. Although there is a correlation between sales and profitability, enterprise performance is not only about being in the right place (strategy) but also about getting there faster and at a lower cost than the competition (operational improvement). The impact of efficiency initiatives rolls directly to the bottom line. It therefore stands to reason that, all else being equal, a continual focus on operational improvement may well be what puts a market leader in that position.

Every enterprise has built structures to organize resources, investments, and focus in order to deliver the greatest value to its customers while simultaneously generating profits for its shareholders. Unfortunately, the alignment of resources rarely, if ever, mirrors the true need. People and resources may well be dispensed based on perceptions of opportunities or simply office politics. The end result is a misalignment of resources. This is called a *resource efficiency gap*. Waste exists when resources are not fully used or where available capacity exceeds demand. To a large extent, this is an issue born of incomplete and inaccurate information at the managerial level. Leaders simply do not know where to add and where to cut.

One way enterprises rid themselves of wasteful spending is through periodic enterprise-wide cost-reduction initiatives. These programs are designed to take a holistic view of an enterprise and identify low-hanging fruit—costs incurred without an equivalent value created. Under the mantle of these programs, resource efficiency gaps are attacked and eliminated. The aim of an innovative enterprise, however, is not to periodically eliminate waste but to continuously reevaluate resource requirements across the enterprise and realign resources to efficiently fill needs—not wants.

In a process-based approach, operational efficiency receives its due respect. The strategic planning process brings to light capabilities required to compete effectively in the future. These requirements spawn operational efficiency initiatives to enhance the enterprise's capabilities. Additionally, one of the core responsibilities of a process owner is to continuously evaluate the processes under his or her ownership and to identify improvements. These improvements may be the development of capabilities or the shedding of unnecessary ones. In this way, the scale slides both ways—for both the growth and the diminution of processes. For example, operational-improvement initiatives might focus on providing strategic flexibility, expanding systems and networks to accommodate growth, or improving the quality or safety of a process, and other times they may jettison unnecessary capabilities.

Over the course of executing multiple enterprise cost-reduction programs, I made an interesting discovery about the management of costs in most enterprises. A significant percentage of cost falls in the sales, general, and administrative (SG&A) bucket, but this line item is rarely the focus of an improvement undertaking. But SG&A expenses are not the only victims of neglect. There are almost always ignored and unloved areas in just about every enterprise. And they are ripe for improvement. Why the neglect? The answer is simple: these opportunities lie in a veritable no-man's land. No one is assigned responsibility for examining this area for opportunities.

Fortunately, in a process-focused enterprise, there are process owners. Whereas the innovation portfolio operates at an enterprise level, work efforts not large enough to be funneled into the innovation portfolio are delegated to process owners. To address these smaller opportunities, process owners are trained to identify and execute small improvement projects. For the most part, these projects need only a pinch of time to see them through to completion. Because flex time exists (and it exists in every enterprise), process owners set improvement goals for their processes—focusing on cost takeouts, quality improvements, expanding throughput, enhancing the skills and knowledge of performers, and expanding the capabilities of the process—and they use available resources to get the job done.

As operational-improvement opportunities come to light, the process owners and other leaders identify and scope initiatives in the same manner as strategic initiatives. As the initiatives are identified and built out, they move to the initiative-management function for execution.

Initiative Management

As initiatives of all stripes are created, management analyzes and prioritizes their launch in any number of ways. Based on my experience, the overwhelming number of organizations have forgotten (if they ever knew) the full intent of an initiative-management function. Initiative management provides not solely a platform for the development, prioritization, and launch of initiatives but also a methodical process for continually reassessing and adjusting the enterprise's efforts to deliver the greatest return on its initiatives. This aim of portfolio management is sadly missing in all but a token few enterprises today. But there are additional benefits when an initiative-management function uses process adjustments to communicate intended improvements.

The first benefit is the crystalline clarity and understanding of the actual improvement initiatives. As initiatives and their respective end state are developed from a straw man to detailed solutions, an initiative's impact on the overall process structure and the expected outcomes becomes unquestionably clearer. Although mentioned previously, it bears repeating that this alone is a tremendous leap forward for every enterprise. No longer is a project team forced to guess intent or translate from leadership directives what an initiative is really supposed to accomplish. With a clear target, the percentages swing in favor of success.

The second benefit is equally important. As initiatives are prioritized and slated for launch, understanding what processes are affected provides visibility to where improvement efforts may stress or overwhelm a particular area. Every team and function has a limited capacity to absorb change during any period of time. By scheduling the deployment of solutions, leaders can prevent the swamping of a particular area and the deployment of conflicting initiatives. For example, if both a strategic and an operational-improvement initiative exist for the same process, the strategic initiative should always take precedence over any attempt to improve the efficiency of the process.

Otherwise, the efficiency initiative may aim to improve a process that no longer exists after the strategic initiative is complete.

In my experience, I have found that one useful tool to visualize process impacts is a heat map of the process structure. Through the use of colors or other denotation, process impacts can be depicted for each process—initiative by initiative. In this way, process owners and other leaders can review the heat map for a process and determine whether there are overlaps in initiative scopes that require investigation and possibly mitigation.

In a process-focused enterprise organized around the four facets, the goal is to laser the resources, energy, and focus of the enterprise's capabilities on areas where improvements deliver the greatest value. Using process as the language of change allows for a very specific diagnosis of what needs to change—and an equally accurate picture of the implications. In this way, investments of money, time, and energy are directed to where they contribute the greatest value.

BENEFITS OF USING A PROCESS-BASED APPROACH

For the past century, big business operated with a traditional hierarchical organizational structure and a collective subconscious expectation that having a seasoned leader at the helm was the way to drive performance. Failures were chalked up as shortfalls of a specific leader or changing market conditions that no one could foresee. But rarely are the structure and organization of the enterprise itself blamed. In the absence of a framework to manage innovation and the supporting structure to enable its execution, leaders and planners continuously reinvent the wheel. Every innovation begins as blank sheet. This adds complexity, delays, waste, cost overruns, and a host of other negatives, but most important, it usually results in an inferior end state. Imagine a different environment where processes hum with efficiency, timely and insightful reports enable leaders to craft game changing strategies, and the infrastructure is in place to execute these initiatives in a timely and methodical fashion. No longer are organizational structure, institutional knowledge, and operational clarity impediments to change; now the foundation is readily available and capable of acting as a springboard to take the enterprise anywhere it wants to go.

To date, the greater business community, academics, and all the armchair quarterbacks have failed to provide a system that continually and effectively innovates. As a result, the core structures and practices inside most enterprises have remained relatively unchanged for decades. A process-based approach provides an alternative to the status quo, releasing the full power of an enterprise to better serve the consumer and increase returns to shareholders. Such a model is vastly superior to conventional approaches in its simplicity, specificity, and ability to deliver immediately. Several of the major differences between the status quo and this new approach include the following:

- Customer insights are captured in multiples of what is provided by existing practices today. The use of feedback loops from frontline associates and the active management of customer-facing processes facilitate prompt reactions to consumer and market forces as they change.
- Strategic planning is less guesswork and based on gut feeling and more a systematic and precise procedure. Strategy translation problems are mitigated as a result of the specificity and clarity of the solution defined in process adjustments.
- Enterprise alignment occurs more naturally because the workforce is knowledgeable about the enterprise's innovation plan as communicated through strategic and operational initiatives.

- Continual improvement occurs throughout all levels and every corner of the enterprise. Process owners take responsibility for improving the efficiency of the processes in their domain. Cross-functional/process-improvement efforts are identified, designed, and funneled through an initiative-management function in a manner that maximizes the benefit to the enterprise.

- The overall enterprise becomes more effective and efficient. Waste, miscommunication, and unclear leadership directives are reduced. These improvements build momentum and excitement as employees on the periphery of innovation activity buy into the methodology. Morale and employee engagement improve, often creating a tidal wave of communal pride.

As with any system or approach that relies on human beings for its execution, initiatives will still occasionally fly off the tracks with a process-based approach. Political power plays, although minimized, still may flare up occasionally and impede forward progress. Strategies forged on faulty information or poor design still may be launched and crash. Tools and methodologies will be misapplied and waste time and money. Associates will still resist adjustments to their world, stalling improvement efforts. In spite of these risks, implementing a process-based approach creates a structure uniquely responsive to change. The question is: Is your enterprise ready to commit to the change? The first step on this path is to recognize who you really are and how you operate—through an exercise to uncover the process structure operating in your enterprise today.

5

Process System: Only Build with an Accurate Blueprint

The late Barry Goldwater was a political icon; he was the Republican nominee for president in 1964 and represented Arizona in the U.S. Senate for 30 years. In 1979, during the Iran hostage crisis, the Senate Foreign Relations Committee, of which Goldwater was a member, met to discuss the possibility of launching a military mission to free the American hostages. As the discussion raged, Senator Goldwater realized that several committee members were unaware of Iran's geographic location. This lack of a foundational understanding hindered the debate on the risks of destabilizing an influential player in such a volatile region of the world. Before the next committee meeting, Senator Goldwater took it upon himself to nail a world map to the wall of the conference room. His aim was to expand the committee's collective awareness of the area's geography to foster discussion on the merits and risks of a rescue attempt.

The map provided the needed clarity, allowing the committee to move forward with a shared foundation of knowledge.

The arena of politics is far from alone in its need for a foundational contextual understanding when debating options. As told in the Digital Equipment Corporation (DEC) example in the Preface, many leadership teams suffer from a fragmented understanding of their organization's capabilities. But this deficiency does not appear to minimize the zest of leaders for spinning the wheel and launching new directives. Year after year, leadership teams craft plans to drive the performance of their enterprises to new heights. Yet, strangely, the plans are launched with minimal consideration of the existing capabilities of the enterprise. In any other realm, undertaking a major endeavor without a solid baseline from which to build would seem absurd. Would any sane contractor build a skyscraper without meticulously drawn blueprints detailing the full range of internal systems and structural components? No regulatory agency would approve a building permit. No reputable subcontractors would consider working on such a project. No bank would finance it, and no insurance company would underwrite a policy on it. Yet this is exactly what happens in the business world on a daily basis. Leaders initiate aggressive agendas to take their organizations to the promised land—but without solid confirmation of their feasibility. Success on the chosen course is often only possible with the arrival of a hero to deliver the impossible and triumph over ambiguity and organizational deficiencies.

An understanding of the current environment supercharges the innovation engine. As is clear from the Goldwater example, a shared view provides clarity for planning, especially in instances where structures, processes, and organizations already exist. It provides ground zero for any debate on opportunities—allowing managers to start with the same background of knowledge and build from solid ground. Refurbishing a home is a parallel with enterprise innovation.

The blueprint delineates the existing structure and the intended addition to be built—using the existing structure as the starting point for planning. In enterprises, the intent is the same. Enterprises proficient at innovation understand their current structures and use them as a foundation from which to build improvements. Starting in a vacuum wastes time, energy, and dollars—factors that may well determine success in a competitive environment.

TOOLS FOR FOUNDATIONAL UNDERSTANDING: THE ORGANIZATIONAL CHART VERSUS A PROCESS-BASED VIEW

When questioned about how their organizations operate, most leaders today pull an organizational chart from their files and point out individuals and their responsibilities. The modern organizational structure is a grandchild of the military hierarchy that became prevalent during centuries of warfare. The typical organization structure is a cascading structure that shows the leader of each business unit and their direct reports. Titles or positions are commonly listed for each individual. Depending on the scope of the organizational chart, additional direct reports and their direct reports may be identified as well. Although loosely tied to functional structures, all the workers are not included, or they are identified in such a vague fashion as to complicate understanding of where and by whom specific activities are actually performed. Although such a chart is effective for planning leadership succession, it is a rather inefficient model to depict the intricacies of a complex business system.

Organizational charts and the functional structures they represent are inadequate foundations for improvement efforts. Not only are they based on people and not activities, but they suffer from two glaring omissions. To start, the customer, the primary reason for the enterprise's existence, is missing from most contemporary organizational charts. Second, most organizational charts fail to specify the flow of work, including all the processes and their interdependencies. Fortunately, an alternative mechanism to depict enterprise operations is available—one based on a well-known way of organizing work activities.

Every enterprise, no matter what size or legal classification, consists of a collection of interconnected and interdependent processes. These processes and their connectivity with external stakeholders are the mechanism through which an enterprise delivers a product or service, generates information, or promotes a cause. The full strata of processes constitute a *process system*. In general, a process system is analogous to the human body. The human body consists of trillions of specialized cells grouped together to form organs and structures (i.e., muscles, bones, etc.). These specialized groupings serve to give us the ability to think, walk, talk, and perform a nearly infinite number of activities. In a similar fashion, an enterprise consists of processes, systems, and people who work together to create products and services.

Consider Olympic athletes and how they maximize their natural abilities to attain an athletic prowess that is far beyond that of an average individual. Year after year, they manage their exercise, eating, and sleeping habits to obtain peak conditioning—putting them in a position to compete at the highest possible level in their sport. In a similar way, enterprises need to plan and build their capabilities to compete in a marketplace. This brings us to the crux of managing a process system.

The challenge in a competitive marketplace is to proactively manage processes—adjusting them to produce exactly what is needed to satisfy customers and driving them to achieve performance goals. Actual performance, be it financial growth or strategic advantage, is largely determined by the efficiency by which an enterprise improves its overall process system to deliver products/ services that most closely correlate with the wants and needs of its customers.

The Enterprise Process System

An enterprise that proactively and appropriately manages not only its ongoing operations but also its future capabilities is destined to become the industry heavyweight. Building such an innovation machine requires as a first step that leaders have a clear and shared picture of their organization's capabilities, allowing them to collectively focus on what needs to change and to commit to making it happen. The question, therefore, is how to create this shared view. Fortunately, there is a construct that provides the benefits of the organizational chart, incorporates customer connections and clearly depicts the flow of work through the network of processes. This model is called the *enterprise process system.*

The enterprise process system includes all the processes in an enterprise—not roles or functions, but processes. Using processes as the base-work structure yields the simplest and most meaningful view of the actual ground-level work occurring in an enterprise. A process system segments processes into logical groupings based on their relationship to each other. In this way, a complete view of an enterprise's operations can be built. This delineation of process begins with the highest level of the process system—the *enterprise process blueprint*—and cascades down to more granular work groupings until it arrives at the lowest unit of work—the *execution of a one-step task.* A standard breakdown of the process system is as follows:

- *Enterprise process blueprint.* The first-level view of the process system is the enterprise process blueprint (also called an *enterprise process map*). The enterprise process blueprint provides a bird's-eye view of operations. As a pictorial representation, the blueprint identifies the core customer of the enterprise and depicts how the megaprocesses work together to produce outputs for a customer. Arrows showing the interaction between megaprocesses and other entities represent the general flow of work. An enterprise process blueprint is shown in Figure 5.1.
- *Megaprocesses.* The second level in the process system is the megaprocess. Megaprocesses in Figure 5.1 include "Sales," "Website," "Marketing," "Manufacturing," and "Supply Chain." With rare exception, the megaprocesses are not singular processes but groups of related processes, represented by the rectangular boxes inside the megaprocesses. When

FIGURE 5.1 Enterprise process blueprint for a manufacturing company.

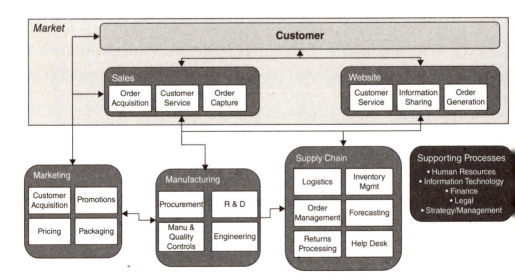

compared with the traditional organizational structure, megaprocesses are roughly equivalent to functional areas or departments. On average, a for-profit company has between 6 and 12 megaprocesses. Nonprofits and other institutions vary greatly in their number of megaprocesses. The boundaries for megaprocesses exist solely to provide management over similar processes and not to segment the design or execution of work.

■ *Major processes/processes/subprocesses.* Megaprocesses can be further broken down into major and minor processes. Major processes include traditional processes such as order acquisition, order taking, order fulfillment, and a gazillion others. Many major processes may be broken down further into processes, although major processes do not always require further segmentation. The major process "Order Taking" may encompass processes that are channel dependent, such as "Online Order Taking," "Sales Force Order Taking," and "Call Center Order Taking." Processes may be further broken down into subprocesses. This further segmentation is appropriate when a process is too large to be managed by one individual or team or when the process entails a focus unique to a certain customer segment or channel. In this instance, a process owner or manager with correspondingly unique capabilities may oversee the subprocess.

■ *Activities/tasks.* Activities and tasks are the lowest increment of work elements identified in most process systems. Activities are the building blocks of processes. Similarly, activities consist of one or more tasks that together serve a specific purpose. Improving the performance of an activity is accomplished by changing, adding, or eliminating tasks. By themselves, tasks are of a specificity that prevents further adjustment.

■ *Procedures.* Often a rough process exists, but because of variations in the inputs or outputs, the work activities to deliver the output vary, and therefore, the process cannot be documented accurately in a step-by-step manner. In lieu of a robust *A-to-B* process, procedures are general guidelines to direct the completion of work. Procedures still add value to the enterprise—the value is just tied to the flexibility in their execution. Procedures are prevalent in customer service and other front-end positions where associates customize their responses based on customer requests.

The process system is not only a representation of an enterprise's operational structure. It also identifies logical work units that can be managed and adjusted—setting up the foundational view for innovation that is lacking in most business environments. That said, it is rarely beneficial to fully design and document every process in an enterprise. The key is to identify the processes that are the primary generators of value in the enterprise. These processes are where time and investments produce the greatest value. However, no level of process should ever be completely eliminated as a candidate for improvement. It is not at all uncommon to find opportunities in seldom-used processes with investment potential that exceeds that of more salient processes. This occurs because the more prominent processes are continuously harvested for improvement opportunities, whereas seldom-used processes are neglected and therefore become fertile terrain over time. The process system's capability to be a straightforward foundation on which to base improvement efforts depends on both its accuracy and leadership's buy-in that it accurately depicts how the enterprise really operates. The first step in uniting leaders behind this shared perspective is to build an enterprise process blueprint.

Enterprise Process Blueprint Defined

As stated earlier, an enterprise process blueprint is a bird's-eye pictorial representation of a process system. It depicts operational components and includes all the major functions/processes in sufficient detail to minimize any confusion. An enterprise process blueprint is a unique view of an enterprise and is extremely useful to manage or innovate an enterprise. In short, an enterprise process blueprint

- Begins with the customer. It is *customer focused*, with identification of the serviced customer segments and the specific connections through which the enterprise interacts with customers. Because the customer is the reason why the organization exists, it is only logical to use the customer as the starting point for the organization's operations.
- Provides an *end-to-end view* of the enterprise, including connectivity with raw material suppliers, business partners, and customers participating in the value-creation process.
- Clearly *identifies the flow of work* in the enterprise. It shows the connectivity between the major functions/processes and how products and services are produced.
- Is *process focused*, with identification of megaprocesses and major processes as well as, on many enterprise process maps, identification of the process owners.
- Provides *simplicity and clarity* of an enterprise's value chain and its supporting processes—creating a model that can be used as a starting point for planning exercises.

Figure 5.2 shows the enterprise process blueprint of a well-known Fortune 100 retailer. This example includes all the major

FIGURE 5.2 Retail enterprise process blueprint example.

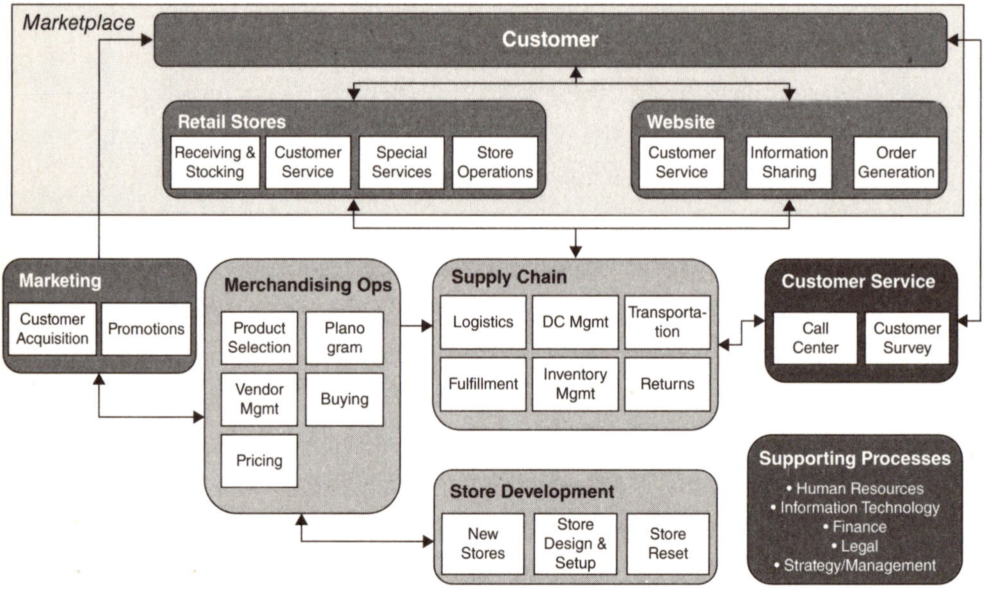

components of a well-developed enterprise process blueprint. It starts with the customer seated at the top of the diagram, it identifies customer touch points, it lists seven megaprocesses with their related major processes, it identifies the primary supporting processes, and it uses arrows to depict the general flow of work.

When viewing an enterprise process blueprint for the first time, many professionals are underimpressed by its simplicity and may even suggest that it is little more than common knowledge. Indeed, the process to create an enterprise process blueprint generates a collective lucidity. But do not be fooled by its simplicity. The enterprise process blueprint is one of the most powerful tools for spreading a shared operational awareness across a management team. For perhaps the first time, the workflows are not just identified but are also implicitly recognized as the avenues of productive capacity that span business units, teams, and departments. After it is fully assembled, much like a puzzle, it is easy to see how the pieces fit. With such a clear view of operations, leaders and managers cannot help but begin to define their vision for the enterprise in terms of the blueprint.

CREATING AN ENTERPRISE PROCESS BLUEPRINT

In stark contrast to the appearance of the end product, the process for creating an enterprise process blueprint is frequently trying and can get messy. Executives and managers are forced to put on paper what up until now has been ambiguous and vague. Misunderstandings and differences of opinion are inevitable. The process itself is one of discovery, and it requires that decisions (or commitments) be made as to how the enterprise really functions. Although there is no right answer, the end result should represent the collective opinion of the leadership team, including the major functional leaders. To reach this endpoint, the creation process always concludes with a final confirmation session—where attendees challenge the enterprise process map, driving to make it as real as possible and resolving any outstanding questions. On occasion, consensus will not be reached because entrenched opinions may end up being insurmountable. While not ideal, this is acceptable. The enterprise process blueprint can simply be labeled as a draft with the discrepancies noted. Over time, differences of opinion will fade as business-as-usual practices are further studied to conclusively resolve any remaining areas of disagreement. And, of course, as the enterprise evolves, the process blueprint will require updates, edits, and additions.

The creation process consists of five steps. Without a doubt, the greatest challenge in developing the blueprint is the forging of consensus across a wide band of leaders. Active listening and careful documentation are necessities for the team tasked with its creation. As the saying goes, the devil is in the details.

Step 1: Leadership Interviews

Building an enterprise process blueprint in anything approaching a timely fashion requires a dedicated team to go where the information resides—inside the heads of leaders, managers, and workers throughout the enterprise. Definitely leverage any previously completed process flows and organizational charts, but recognize that the intended outcome is a collective view of operations. Achieving this view entails the engagement of leaders and managers—and the extraction and accumulation of their perspectives. The knowledge gained during the process is nearly as valuable as the final deliverable.

To begin, start at the highest levels of senior leadership. From their positions atop the organizational pyramid, these individual have the broadest view of the enterprise's operations. Their viewpoint is helpful not only to get a rough outline of the operational areas but also to identify the next layer of leaders to interview. Aim to walk away from these initial discussions with a draft of the megaprocesses and loose associations of the major processes. This view will be enriched by the inclusion of the perspectives of the second tier of leaders. Because most enterprises are functionally based, odds are that the second round of interviews is with departmental leaders—usually vice presidents and directors. Their insight is critical to correctly associating major processes with the greater megaprocesses. Also be sure to include a healthy number of leaders from outlying operational units—especially those in geographically diverse locations. Operationally independent leaders often possess unique perspectives on how the enterprise operates. As the overall picture unfolds, refrain from accepting any single individual's input as gospel. Everyone is a victim of perspective to some extent. Every answer is colored by the interviewee's unique vantage point and situational circumstances. Although honestly provided, their responses are opinions. Trust but verify.

When conducting informational interviews, it is a mistake to expect anyone to fully understand his or her processes. Many concepts that are basic to process experts are foreign to other folks. For this reason, the ability to collect the information and accurately document a process system depends on the interviewer's prowess. Because the aim is to capture information of an equivalent depth and quality from each interview, I recommend using a structured approach with a scripted set of interview questions. The following list provides sample questions to include in an interview guide:

- What are the major deliverables of your area? Are there other deliverables your team provides?
- What are the inputs to these deliverables? From where do you obtain these inputs?
- Who are the performers of the process? What percentage of their time do they spend on each process?
- Where are the outputs delivered?
- How is success measured? Metrics?
- Who are the business partners in creating the deliverables?
- Which deliverables, processes, or functions are most critical to your business function?

Although some of the answers may not make it to the page, they are helpful to pull information out of the interviewees. And there are multitudes of other questions that are applicable. Consistency is the key to crafting an accurate process system. Take the process discussions to a predetermined depth of detail. Not doing so creates the potential to misinterpret the scale and import of a particular process in relation to others.

Step 2: Initial Draft

After completing these two rounds of interviews, take that first shot at documenting a draft of the enterprise process blueprint. In this endeavor, do not let perfection be the enemy of progress. Just get the facts on paper, and allow them to morph into form through continued investigation, discussion, and debate.

- Start with the customer(s). Draw a box to indicate the customer(s) of the enterprise. Depending on the mission of enterprise, the customer may be a student, beneficiary, or constituent.
- Identify the touch points and channels between the customer and the enterprise or the enterprise's partners (e.g., distributors, retailers, or other distribution partners). Include all customer touch points on the blueprint, even when they are not directly connected with part of the company. Any intermediary between the customer and the company is a critical component of the process system.
- For each of the touch points in the preceding step, draw a box to identify the process/team/organization interacting with the customer. The exact name is not important at this point—just a specific descriptor that leaves no question as to its identity. If there are multiple touch points, list them all. Do not be overly concerned with accuracy. This is an iterative process, and the blueprint always changes as more information becomes available.
- One caution: always remember that this enterprise process blueprint represents the current state. At this point, it is not uncommon for roles or connections to be poorly defined or not defined at all. This is one of the early benefits of creating an enterprise process blueprint—shining a light into the dark recesses of operations and identifying glaring deficiencies or inconsistencies.
- Below the customer touch points, draw out the core value chain for the enterprise. The core value chain usually includes one or more of the following processes:

- ▲ Acquire/develop customers (also may be a supporting process)
- ▲ Procure raw materials or components of the offered product/service
- ▲ Manufacture/obtain product/service
- ▲ Sell product/service to consumer
- ▲ Deliver product/service to consumer
- ▲ Service the customer after purchase

- ■ Draw a box or a series of boxes to identify the core value chain. When processes overlap or work in tandem to deliver value, only a single box is necessary.
- ■ Identify and draw the links between the core-value-chain processes and the previously identified customer connection points. Use arrows to show the general flow of work as it passes through the organization. The processes identified at this point are the initial megaprocesses.
- ■ Identify any other major components of the enterprise. Draw boxes to denote these groups. Draw lines to indicate their connection to other processes.
- ■ List the supporting processes, such as human resources, finance, accounting, and information technology. Note each of them in a single box for supporting processes. Because these functions/processes generally support the core-value-chain processes (as well as each other), connecting them to other processes would introduce an unnecessary level of complexity to the blueprint. There is an assumed linkage with the other processes/functions.
- ■ Take a step back, and reevaluate the boxes. Consolidate boxes that serve a common function or play a similar role in the enterprise. The remaining boxes are the megaprocesses. Most enterprises have between 6 and 12 megaprocesses, although this number will climb in diversified conglomerates.

Before publishing a draft of the enterprise process blueprint, take a moment to review the interview notes to ensure that nothing was missed. Compare the enterprise process blueprint with the enterprise's current organizational chart. Confirm that all the functional areas are accounted for directly or that they logically fit somewhere if not explicitly noted. On occasion, a process that is significant enough to warrant inclusion on the blueprint will be missed initially. If necessary, reach out to the sources of the notes to reconfirm their responses and to identify the appropriate way to include the process on the blueprint. As mentioned previously, depending on the vantage point of the interviewees, their perspectives on specific processes may differ. When the answer is not immediately available, create a list of questions and points of clarification to resolve at a later time.

Step 3: Distribute the Initial Draft and Collect Feedback

At this point, a reasonably good draft of the enterprise process blueprint is available—although there may be gaps. The next step is to share this draft with the next organizational layer—specifically individuals closer to the work. Although titles vary across enterprises, this group of interviewees includes directors, managers, process leaders, and other individuals who oversee the ongoing everyday work.

Although some enterprise process blueprints only identify the megaprocesses, fully vetting the major processes is a good method

---

Special Note: Because most enterprises today operate using the traditional hierarchical organizational structure, the initial enterprise process blueprint as a current-state reflection will depict megaprocesses and other process boundaries that break end-to-end processes into functional boundaries. As an initial step, this can rarely be avoided. The key principle here is to understand the existing state of the enterprise in order to manage end-to-end customer-based goals. Over time, the enterprise can be reorganized to more closely align the core value chain with end-to-end processes. Figure 5.3 shows the enterprise process blueprint for a small savings and loan. In this enterprise, the depository, loan, and investment offerings are each overseen by megaprocess owners. However, many of the individuals engaged in the work of these megaprocesses are located in the supporting processes. For example, the print shop's employees support all the product lines in the production of customer statements.

FIGURE 5.3 Enterprise process blueprint for a small savings and loan.

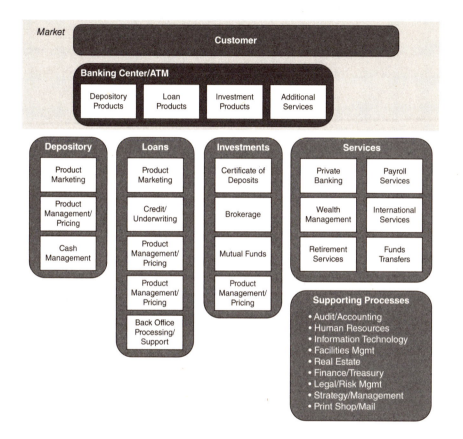

to confirm the accuracy of the megaprocesses. Every process that is performed regularly should be housed in a megaprocess or be called out as a supporting process. Frequently, the megaprocesses need to be redefined to accommodate the full assortment of processes. This round of interviews provides a forum for those closer to the ongoing work efforts to contribute their knowledge and insight to the blueprint. By doing so, they validate the previously gathered information and plug any gaps in the awareness of senior leadership of operational details.

This second round of interviews is also a good time to ask the functional and process leaders to identify supporting processes. On rare occasions, supporting processes are part of the core-value-chain process (e.g., technology processes for a software company). When this occurs, it may be helpful to list the process twice—once in the core-value-chain process and a second time in the supporting-processes box.

As the second round of interviews is completed, return to the interview notes, and begin updating the enterprise process blueprint. Except in very rare instances, there will be differing opinions on how the enterprise operates and where specific processes operate. Add any unresolved questions to the previously created list. Once the newly gathered information is incorporated into the blueprint, distribute it to all participants along with the list of outstanding questions. And then let it sit for a while—at least a week—before holding a final confirmation session. In this interim period, many of the participants will naturally investigate and resolve the outstanding questions.

Step 4: Confirmation Session

With a solid draft in hand, it is time to conduct a final confirmation session to close the door on the outstanding questions and to obtain confirmation. Schedule a time, and invite all the senior, functional, managerial, and process leaders to review the current iteration. Be forewarned, this validation session is rarely a rubber-stamp exercise. The goal of this meeting is to conduct an exhaustive review of every part of the blueprint for accuracy and to incorporate revelations sparked by a collective review.

During the session, introduce the outstanding items for discussion and resolution. If the group fails to reach a consensus on any of the questions, move forward without complete agreement. When disagreement lingers, appoint a small team to resolve the outstanding questions through exploratory visits to observe the process firsthand, through informational interviews with key performers, or via another discovery method. In my experience, many of the outstanding questions are naturally answered with the passage of time—simply because the team sees how the process is really performed.

Step 5: Make It Available

After the edits from the confirmation session are complete, distribute the enterprise process blueprint far and wide. Pin it on bulletin boards, and make posters of it for common areas. Share it with the full universe of employees. Make it an agenda item for staff meetings, executive retreats, and departmental meetings. The enterprise process blueprint is the foundational view of the operation. Let it become the reference point for brainstorms, discussions, and debates.

After publication, additional questions will surface that require the creation team to reconvene and create updated versions of the blueprint. Iteration is always required. It bears repeating: most leadership teams do not have a firm understanding of the operational details of their enterprises. Expecting 100 percent accuracy the first time is unrealistic.

The creation of an enterprise process blueprint is an arduous and exhausting task that requires a high degree of persistence to push through ambiguity, navigate executive egos, and motor through passive resistors. To simplify the creation, there are rules of thumb to hasten the process, make it less cumbersome, and improve the quality of the end result.

Rules of Thumb for Creating an Enterprise Process Blueprint

- When identifying the process system, avoid the use of names of current or planned departments or business units (with the exception of supporting processes). This reduces the chance that leaders will play political games to preserve or grow their fiefdoms. Additionally, the standing names rarely convey the true function or process being performed. Start with new names, and discard the baggage associated with the current nomenclature.

- Although it is extremely difficult to do so, resist noting owner's names on the blueprint. The people part of the enterprise comes later. An enterprise process blueprint is a view of the current state that may well differ from the current ownership. Specifying ownership at this point convolutes the creation to be about mapping the organizational chart to the process blueprint. This pollutes the logic and structure of the blueprint. For now, focus on the processes—ignore the people.

- Recognize up front that core value processes span multiple functional departments. The goal is not to restate functional separations but to view the organization as a collection of processes. When identifying boundaries between megaprocesses or processes seems to be impossible, view the processes from the perspective of their outputs. If the processes work in tandem to create an output, this means that the processes are linked. All the steps and activities leading to the creation of this output can be grouped together.

- Allow sufficient time for reviewers to critique the enterprise process blueprint. For most leaders, the blueprint is a radically new view of the enterprise, and it requires time for acceptance. Rushing individual reviews increases the likelihood that the end result will mimic the deficiencies and boundaries of the current functional organizational structure. This occurs because when under pressure, folks will toss up their hands and revert to the familiar. A better approach is to set loose deadlines and maintain an active dialogue with reviewers to answer questions and move the team toward eventual adoption.

■ When building an enterprise process blueprint, it is common to hear requests to build the blueprint to reflect a future state. Deny this request. Although a future-state map may be useful as a target, focusing on it as the initial exercise unnecessarily introduces confusion, turf wars, and political games. Such a process pushes leaders to define a future state without a clear understanding of the current state. Can you plan a road trip without knowing your starting point? Although a future-state view is theoretically the target, a blueprint of the current state is necessary to understand the existing structure and its capabilities and to determine the feasibility of the future state.

■ Use an objective party (external consultants are a good option) to manage creation of the enterprise process blueprint. An objective and experienced team hastens the delivery and heightens the quality.

■ Begin using the blueprint immediately to frame improvement discussions—even when it is still in draft form. This builds familiarity and momentum toward eventual acceptance of the blueprint as the foundational view of an enterprise's operations.

It bears repeating: developing an enterprise process blueprint is messy and frustrating. There may be heated discussions. Your hands will get dirty. Stick with it. The blueprint is the beginning of a new paradigm for managing and growing the enterprise. Although the act of creation may be frustrating and bogged down by continual challenges, the eventual benefits are manyfold. When complete, an accurate enterprise process blueprint is a launch pad for all types of planning activities. The blueprint becomes a staple of leadership discussions—continuously referenced and carted from meeting to meeting.

FLEXIBILITY OF AN ENTERPRISE PROCESS BLUEPRINT

Despite the name, an enterprise process blueprint does not have to depict the enterprise level. Process blueprints may be built for divisional, business unit, product/service family, or departmental levels. The approach and end result are the same. Identify the highest-level processes, and drill down to the process flow level. Likewise, the best process-focused enterprises drill down below the enterprise process blueprint to the basic process flow level.

Along the same lines, an enterprise process blueprint may depict a slice or the full scope of an enterprise's structure depending on its size and complexity. For-profit corporations fall into one of three categories: single-product-family companies, conglomerates, and diversified companies. Nonprofits and other institutions have varying structures but generally will map to the base-level enterprise process blueprint (single-product-family companies). There are exceptions, however. Some large nonprofits structurally resemble conglomerates or diversified companies. All enterprises, regardless of their legal status, can be mapped to one of the three corporate categories just identified.

Single-product-family enterprises are the easiest to understand and map. They sell a single product or group of products to a limited customer group. Their operational structure is relatively straightforward and can be depicted in a single diagram. All the different elements of the enterprise support the market strategy for a single customer group. The single product family is the base level of the enterprise process blueprint—the view used to provide instruction on enterprise process blueprint creation.

Conglomerates and diversified companies may manufacture a multitude of products for the same customer base or multiple customer segments. Customer segments may be differentiated by product usage, customer purchasing process, or even geography. The primary difference between conglomerates and diversified companies is that conglomerates service different customer segments through stand-alone business units. Any centralized corporate structure exists only to manage the portfolio of businesses—often buying or selling business units. Diversified companies deviate from the conglomerate model in that the business units share supporting processes. The corporate structure atop a diversified company is largely shared service functions (e.g., finance, marketing, human resources, and information technology). On occasion, a diversified company merges other elements related to the servicing of a specific customer segment (e.g., merged sales force or a shared distribution network). Conglomerates and diversified companies are somewhat more challenging to depict with an enterprise process blueprint. As a general rule, an enterprise process blueprint should be created for each distinct product/service offering.

A *conglomerate* may require multiple enterprise process blueprints to represent its full scope of operations. Each map would depict a distinct business unit with its unique customer segment. Often an additional page is used to show the relatively abbreviated corporate structure—usually just executives and strategy/merger and acquisition teams.

A *diversified company* is the hardest of the three to depict in an enterprise process blueprint. Several possible depictions of this operational structure have evolved. The choice of the three depends on the leadership's intent.

The first method is to depict each business unit in the same manner as the conglomerate. This approach leads to duplicative representations of shared functions on the blueprint for each specific business unit. If the leadership thinks of the individual businesses as components of a portfolio, this blueprint is the best choice. It allows for the leadership team to quickly identify salable businesses and the requirements to make each business self-sufficient.

A second method is to not denote the shared functions on the enterprise process blueprint for each business unit but rather to place the shared functions on their own blueprint with the business units as the customers. This method also can be used to incorporate key business partners of an enterprise. This choice is appropriate if the leadership team intends to operate the businesses independently but centralizes shared service functions to capture efficiency benefits.

A final method is to depict the business units and the shared functions in a single enterprise process blueprint—although size and complexity frequently may make this approach impractical. This method is appropriate if the enterprise is looking to connect all the unique businesses together and reengineer the overall operational structure of the enterprise.

All three methods are correct and appropriate. The best choice is to understand the perspective of the leadership team and how its members view the operational linkages between the business units. Then build the enterprise process blueprint using the approach that best represents leadership's intentions.

In an effort to make a process blueprint as user friendly as possible, formatting such as shading and colors can be used to identify core processes versus supporting processes or to identify customers, vendors, suppliers, and other partners. Other opportunities to format the enterprise process blueprint to quickly provide additional information include the following:

- Use the sizes of the boxes on the blueprint to represent head count, budget, number of managers, or another determinant of size. An interesting exercise is to create multiple enterprise process blueprints with process boxes corresponding to elements such as head count, expenses, or budget and overlay them to see where there are inconsistencies.

■ Use different colors to denote the status of process efforts. For example, a red process may be in a critical condition and require immediate attention. Colors also may represent the current functional boundaries or where specific leaders have responsibility. In the same way, colors may be used to denote organizational boundaries between business partners, customers, and other groups.

■ Identify or list the leaders or process owners on the enterprise process blueprint. This provides a quick reference as to the individual(s) with knowledge about and responsibility for any specific component of the enterprise process blueprint. Again, always delay adding names to the blueprint until after the initial draft is completed and confirmed by the leadership.

THE HIERARCHY OF PROCESS

The enterprise process blueprint is a major step on the path to building a foundational understanding of an enterprise's operations, but the view it presents is at a high level and significantly distanced from where the proverbial rubber meets the road. The process level is where the real work gets done. Therefore, a more granular view of work activity is necessary if we want to understand the everyday performance of workers. This is important because improvement efforts require analysis, design, development, and deployment of solutions at the level where the work is performed.

For example, the supply chain is where enterprises manage the processes for receiving raw materials from suppliers and distributing finished goods to customers. When analyzing the cost structure of their operations, leaders often ask for costs to be allocated at the item level in order to price them appropriately. The standard approach is to identify all costs and allocate them to the product using activity-based costing or a similar allocation method. Unfortunately, this approach is fundamentally flawed and results in incorrectly priced products. For example, outbound delivery costs are incurred by truckload and include the costs to load, operate, and unload a truck. The truckload is the cost driver, and efforts to reduce costs must be focused on minimizing the number of truckloads or ensuring that a truck is fully cubed out (i.e., operating with the maximum allowable load). It is not logical to think of these costs as being incurred at the end-product level because this allocated cost would fluctuate widely based on the count of items in a truckload. In other words, the real cost of the item should not be affected by how full or empty the trucks are. Unless a cost is directly attributable to an item, it should not be incorporated as part of that item's cost. For the exact same reason, improvement efforts need to be focused at the appropriate level. Although it may seem like an overly simplistic rule, this mistake occurs with an astounding frequency in the corporate world.

Extrapolating from this example, making substantial improvements requires coordinated action across multiple levels of the process system. The starting point may be the enterprise process blueprint (if only to create a shared view), but real change occurs across the megaprocesses, major processes, processes, subprocesses, and even the singular activities of workers. Thus, although the enterprise process blueprint is the appropriate starting point, additional process views are needed—including digging down to the steps an employee takes to complete a simple task.

MEGAPROCESSES

Most change of a strategic nature occurs at the major process level, process level, and subprocess level. Figure 5.4 is a user-friendly depiction of a megaprocess and the relationships between the major processes associated with it. Whereas there usually 6 to 12 megaprocesses on an enterprise process blueprint, the number of major processes per megaprocess is usually less (4 to 8 major processes per megaprocess). A complete megaprocess visual includes the inputs, outputs, metrics in place, and major processes and their relationships.

This view identifies the inputs, outputs, and primary metrics for the megaprocess. Again, be wary of metrics. Functional and managerial-level metrics create a risk that improvement efforts will be focused on a part of the process system, occasionally to the detriment of the whole system. Metrics at the process level are "safe" only if the full end-to-end process exists inside the process.

FIGURE 5.4 Megaprocess example.

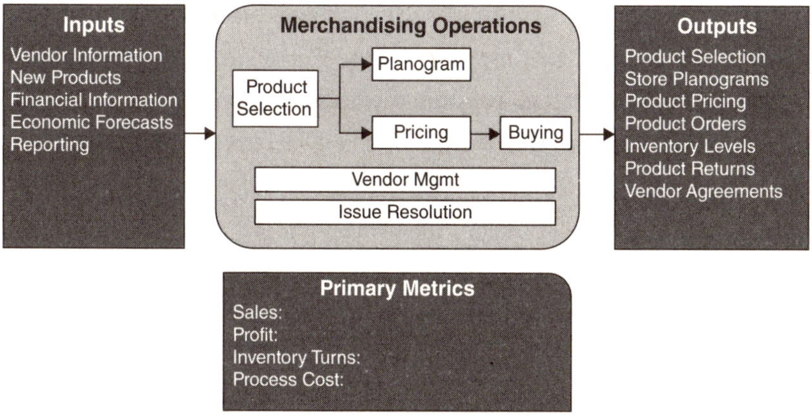

Inputs	Merchandising Operations	Outputs
Vendor Information New Products Financial Information Economic Forecasts Reporting	Product Selection → Planogram Pricing → Buying Vendor Mgmt Issue Resolution	Product Selection Store Planograms Product Pricing Product Orders Inventory Levels Product Returns Vendor Agreements

Primary Metrics
Sales:
Profit:
Inventory Turns:
Process Cost:

For example, the full process cost in merchandising operations is the total cost to buy a product, put it in a store, and sell it to a customer. If this metric were to include only costs from a specific area—such as the cost from the vendor—the process owner might unintentionally push costs elsewhere—such as the store. For example, a vendor may not fully assemble a product but rather deliver it to the store, for the store associates to complete the final assembly. Although this move logically lowers the merchandising costs (i.e., costs to assemble the product are removed from the supplier's cost structure), it adds store cost (i.e., the cost for associates to assemble the product). This is a cost transfer, not a cost reduction. More important, the relocation of assembly may well increase total costs because store associates are less skilled in assembling a product than the supplier's workers. And there are potential hidden costs. Assembling products distracts store associates from their primary duty of taking care of customers and may even increase the number of customer-service issues. Because the tracking of a metric spurs action (i.e., what gets tracked gets done), it is not advisable to track metrics for a limited portion of an end-to-end process.

With completion of the enterprise process blueprint, the megaprocesses are usually fairly well scratched out, although perhaps not in a form conducive to communication. The first step is to identify the processes inside the megaprocesses. This activity confirms the megaprocesses and adds further clarity to their boundaries. Start with the core-value-chain megaprocesses because they are historically the most reviewed and understood processes. If they are not identified during the process of building the blueprint, start by listing the major processes in each megaprocess. Probably the most expeditious way to identify the major processes is to interview functional leaders, especially at the director and manager levels.

When depicting megaprocesses, a good goal is to identify the major processes responsible for creating 80 to 90 percent of the

deliverables. The number of major processes/processes per megaprocess will vary widely, especially when documented the first time. As
a rule of thumb, most megaprocesses should include four to eight
major processes.

Once the megaprocesses and major processes are fairly well
defined, a deeper dive into the major processes provides further
understanding and clarity. Major processes can be deconstructed
into a number of processes that are often called *minor processes*
or *subprocesses*. The relationship between a major process, and a
minor or subprocess varies. A process may be a subpart of a major
process or related to the major process by a number of commonalities, including shared performers, scope, or functional focus.
There are several reasons for breaking a major process into lower-
level processes:

- Major processes that work across multiple product channels
  or customer segments require a different approach (e.g.,
  subprocess) for unique channels or customers. For instance,
  variances of a process, may perform a similar function
  across customer segments but have unique elements specific
  to the customer.
- The major process can be too large to be managed by
  one individual and may require a team approach instead.
  This is accomplished by breaking the major process into
  subprocesses or minor processes.
- The major process may be so complex as to require
  additional owners, often with specialized skill sets. This
  differs from the preceding segments in that the process is
  not broken down into manageable chunks but rather that
  process responsibilities differ across the complete process.
  In effect, the full process is actually an aggregation of several
  interlocked processes.

After the major processes are identified, the next step is documentation. In my experience, process flows are the most accurate depiction of how work is completed. Although useful, the enterprise process blueprint and megaprocesses are artificial constructs created to identify the relationships between processes and to determine the appropriate oversight of the process system. By themselves, they do not represent the level where work produces value. Process flows are the base level documentation for improvement activities and are appropriate to depict major processes, processes, and subprocesses.

PROCESS FLOWS

Process documentation facilitates the acquisition of knowledge around the work to accomplish a business goal, including the inputs, outputs, performers, and actual process steps. Over the years, process documentation evolved from a simple list of steps, to linear depictions of activities, to swim lanes, and for Lean practitioners, to value-stream maps. Regardless of format, a process flow models the work activities that are executed repeatedly by workers. It is the level of documentation used by reengineering and process-transformation teams to understand the current state of a process, design a future state, build it, test it, adjust it, train workers on it, and then deploy it.

To capture the information to build process flows, experienced practitioners use multiple approaches. The most commonly used practices (and arguably the most effective) are informational interviewing and direct observation. Informational interviewing is the fastest way to collect a large amount of detail from a variety sources. It targets individuals with specific knowledge in order to quickly identify the basics and then bounces the findings off individuals who provide input to the process or receive output from the process.

By aggregating these individual perspectives, the most accurate view of a process can be captured.

Informational Interviewing

To increase the likelihood of success, a team or individual with process expertise is the best resource to build process flows. However, experienced and objective process experts may be hard to find in the general workforce. When such experts are not available internally, external consultants are an alternative source to conduct interviews and document processes. To ensure consistency in the level of detail and completeness of the documentation, a simple five-step approach guides the capture and enhances the accuracy of process-flow information. The steps need not be completed in the same order as presented, although there is a bit of logic to the flow. As each step is performed, new information invariably will come to light that, in turn, requires adjustments to previously captured information. As with most process activities, iteration improves the accuracy of the documentation. For the process under investigation, the interview should accomplish the following:

1. *Identify the customers.* Start by identifying the process's customers. Do not forget that customers can be internal, external, or, in some instances, business partners that straddle the line between internal and external. Also be sure to look beyond the prominent customers and identify the less visible beneficiaries of the process. Uncovering these less obvious customers may be tricky until the process is studied step by step.

2. *Identify the outputs.* Knowing the customer jump-starts identification of the outputs. Talk with each customer or customer group, and ask what they get from the process. Is it a product, service, information, or something that is

solely a component of a much larger end product? Once the outputs are named, it always helps to dig a bit deeper to capture the key attributes of the output. In effect, identify why the customer wants the output, such as quality, cost, availability, or customization.

3. *Identify the inputs.* With the outputs clearly identified, a path—although perhaps not initially visible—leads back to the inputs of the process. At this point, the question of process scope is front and center. At what point does the process start, and where does it end? In many cases, improving a process requires the incorporation of activities outside the currently defined process boundaries and, with increasing frequency, outside the enterprise itself. As the boundaries of an improvement effort expand, the potential to achieve significant results increases, and the risk that the improvements gained are not offset by unforeseen impacts elsewhere is minimized. To identify the inputs, a good tactic is to focus on the parts of the larger process that are inside the enterprise. Follow the chain of the activities back to where the inputs were acquired. Inputs take many forms, including raw materials, knowledge, or even a signal to start the process. However, the input list often needs to be refined as further clarity around the process is gained. Iteration is again the driver of accuracy.

4. *Identify the suppliers.* With a list of inputs, the suppliers can be identified through their association with specific inputs. This step is arguably the easiest part of a process-documentation exercise. Every input has either an internal or external provider. The biggest mistake in identifying suppliers is to overlook inputs to the process, such as information and knowledge. Without knowing what to produce or when to produce it, the process might not start.

5. *Walk through the process step by step*. At this point, the
process elements that were previously documented provide
a foundational perspective that resembles a loose strawman
of the process. However, further detail is needed. To get an
actual step-by-step view of a process, go to the performers
who execute the process on a daily basis. During these
interviews, start with the receipt of the inputs, and walk
step by step through the process until the arrival of the step
to transfer the outputs to an end customer. Identify every
activity, and list them in the order of their execution. At
each step, ask whether the process follows the step or there
are decisions or alternatives that potentially add additional
steps. Use good judgment. If an alternate path rarely occurs,
it might be possible to ignore it and treat future occurrences
as exceptions. If the alternative path occurs with regularity,
however, it needs to be documented in the process flow.
Document the steps, the order in which they occur, and
the major alternative courses. When this investigation is
performed in a methodical and diligent manner, the results
are an accurate reflection of the process.

Direct Observation

Once there is an initial draft of a process, direct observation is useful
to confirm the draft and identify exceptions. What is initially built
through interviews may well differ from what normally occurs. This
differential exists because most individuals have limited exposure to
the full scope of a process or because their responses reflect personal
prejudices. Direct observation, as the name implies, entails gathering
information firsthand—by either watching the process performers
execute the process or by actually participating in the process.

To ensure that the process is viewed during standard execution
instead of abnormal periods, multiple observations should be

conducted at different times and under different circumstances. Submerse yourself in the process, examining the deliverables at each step in the process and working back to the elements of work required to create each deliverable. As your observations identify adjustments and changes to the initial process documentation, circle back with other performers, functional managers, customers, and business partners to confirm and clarify. The best process documentation is the result of numerous iterations that account for all types of variables (i.e., temporal, situational, etc.).

Exceptions: Is an Initial State Appropriate?

Especially when working with startups or new business lines, the processes under discussion may not be fully operational, making it impossible to document a current state. When this occurs, forego the current state and instead move directly to an initial state. An initial state dispenses with the gaps between the true current state and delivers a complete process by assuming that the missing elements—activities, tasks, inputs, or outputs—already exist. By making this jump, an initial-state process operates as a foundation on which to base management and improvement efforts. As a general rule, the initial state should never be a stretch. Rather, it should be minimalistic and incorporate only the basic requirements of the process. It is merely a short-term replacement for the current state and represents the simplest process to produce the desired outputs using only available and easily obtained resources. However, if large gaps exist between the current state and the initial state, the initial state becomes in essence a future state. A future state, unlike an initial state, requires a significant investment in resources, time, and energy to get it up and running.

Whether the end result is the current state or an initial state, the value in documenting processes is to deepen the organizational knowledge, create a foundation for innovation, and drive consistent performance. For core processes and the most salient processes, documentation often generates awareness of strategic differentiation opportunities. Thus, if documenting processes enables innovation and consistent performance, why not document every process in the enterprise? Forget it—it would be a tremendous waste of time and resources. Any team anointed with the responsibility to document every single process in an enterprise travels on an unending highway. The sheer number of processes is beyond count in most enterprises. The question, therefore, is, What processes should be documented? Should all processes that are actively managed be documented? If so, how do you distinguish between processes that should be managed and the rest of the field?

To Document or Not to Document?

The answer: not all processes can or should be documented. Processes with the greatest opportunity for improvement—strategic or operational—are the obvious priority for documentation. These processes usually reside in the core value chain. But do not ignore the seemingly insignificant processes because they may hide huge opportunities for improvement. And then, it is not uncommon for an "unloved" process to be core to future strategic requirements. Aside from the most frequently executed processes, only document processes when there is a benefit to doing so. That said, the lack of documentation for a process does not mean that it should not be reviewed, improved, and managed. In fact, quite the opposite is true. Any element of work that is consistently repeated demands a degree of attention at least equal to the value it generates. For many processes, though, documentation is either problematic or simply consumes more effort than it is worth. These types of processes fall into one of the following buckets:

- Processes with varied inputs, process steps, outputs, and measures of success—especially when variances in the process contribute to the value delivered. Some good examples are at the front line where associates interact with customers daily. The individuals interacting with the customers must adjust to satisfy the individual customer's needs and expectations. There are no hard and true processes to account for every type of customer interaction. In order to ensure that the associates perform in a manner agreeable to the enterprise, the associates are trained to follow general guidelines or procedures, such as basic rules for resolving customer issues. Roles lumped into this category include store associates, sales personnel, customer-service functions, and the management of those functions.

- Processes where standard operating procedures, standard practices, or work guidelines are sufficient to instruct and train the performer; these activities are not overly complicated, and some variance is acceptable. If variation in the process makes it challenging to document, take the easy route and craft procedures or guidelines. An example of this type of process is the stocking of a specific product on a shelf, packaging varied products for delivery, or transportation activities such as driving delivery routes.

- Processes that are rarely used and have low value/cost are not worth the time to document or manage. Rules are often acceptable to govern these processes. Other times, these activities are left to the performer's judgment. Examples include associate-level activities outside value-added work efforts, for example, travel around the office or disposing of office trash. When execution of these processes falls outside a reasonable norm, corrective action or procedures may need to be instituted.

▪ Corporate roles that require flexibility, planning, and various forms of execution. Many leadership and corporate positions operate with significant variability on a daily basis. These positions are managed (if they are managed) through skill set and knowledge development.

For regularly executed processes that you choose not to document, it is still a good practice to routinely reexamine them to identify any deficiencies or discover latent opportunities. Many processes deserve attention and improvement only when they are executed outside of acceptable performance levels or they conflict with the enterprise's core values. The intent of process improvement is to focus enterprise resources armed with the right tools on the work activities where the most value can be created. The use of a process-improvement tool such as Six Sigma (a methodology focused on eliminating up to six standard deviations of defects) is unnecessary and wasteful when applied to a process requiring less precision in its outputs. Always use the right tool to get the desired result.

IMMEDIATE BENEFITS OF IDENTIFYING A PROCESS SYSTEM

The activities to document a process system breed widespread awareness of how an enterprise operates. It originates with the individuals who participated in the discovery process but gradually expands to a much larger audience. Employees at all levels begin talking about operational adjustments through the lens of process. As more employees grasp the process parlance, the process system becomes foundational to planning activities.

And this brings us to perhaps the most immediate benefit of a documented process system. With large-scale improvement efforts, a traditional enterprise struggles to identify how simple changes will ripple through the interconnected processes and teams. In lieu of an operational blueprint today, project teams commonly perform a carousel of informational/sharing sessions with different business units and leaders—sharing the intent of an initiative while simultaneously gathering needed details. Many initiatives get jarred off the tracks in these sessions because the directives are open to interpretation, and session participants seek to put their imprint on the initiative's direction. For a process-focused enterprise, initiatives are communicated in process terms—framing leadership's intention in the jargon of ground-level work activity in order to minimize misinterpretations. As the process system becomes an accepted way to communicate enterprise change, it begins serving many roles, including

- Creation of a framework for communicating how the organization operates.
- Clarity as to the interconnectivity of processes.
- Identification of operational areas where knowledge and other capabilities exist.
- Creation of an operational foundation on which to base strategic and efficiency initiatives.
- Identification of opportunities for collaboration on initiatives and other improvement efforts.
- Mitigation of the risk of localized improvements having a negative impact on the overall process system.

The process system is only the beginning of a process-focused enterprise—a new way to view, discuss, and debate the operations of an enterprise. But process documentation is just paper—a collection of snazzy diagrams with descriptive titles that offer guidelines for completing work. The real energy—what sets the production wheels spinning—in any enterprise is the people who perform the processes.

6

Process Governance: The Electrician Performs the Electrical Work

E mployees are the kinetic element of every enterprise—that irreplaceable energy responsible for transforming ideas into valued creations. The extent to which employees are organized and directed is perhaps the most significant determinant of success for any enterprise. *Processes* are how work is organized; *process governance* is how employees are assigned to perform processes.

The primary goal of process governance is to put the right people (as determined by their skills, knowledge, abilities, and interests) in roles where they generate the greatest possible value. The words comprising this definition were chosen with the utmost care. Positions are the remnants of the modern organizational chart. Their usage implies that work is divided into discrete, manageable chunks that workers consistently execute for an indefinite period. But work requirements are not static. Customers want higher-quality products at reduced prices, governments approve new regulations that force the adoption of new operational practices, and competitors earnestly probe for ways to steal market share. To be successful over any period of time, an enterprise must be adaptive to the realities of the moment. This requires a flexible and nimble workforce—adjusting employees' responsibilities to capture advantages as the market moves.

Since the dawn of time, debate has ranged as to the optimal method of organizing people to provide sufficient food, shelter, clothing, and security. Perhaps no arena more influenced this debate than that of warfare. As burgeoning civilizations grew and consumed resources, conflict was inevitable. Cities and states armed their citizens—some for protection and others with more nefarious intentions.

As the Roman Empire rose to prominence, a critical determinant of its battleground success was the ordering of its soldiers into operational units—with a cascading structure of leadership. The highest grouping of soldiers was the *legion*. Legions consisted of roughly 5,400 soldiers, which were further divided into *cohorts*. Every legion was comprised of 10 cohorts—with one cohort made up of 800 soldiers and the remaining nine cohorts made up of 480 soldiers each. Cohorts were further subdivided into *centuries* of roughly 80 soldiers. The last division was the *unit*, a group of eight soldiers. Throughout the army's organization, leaders were appointed to oversee each level based on their social standing and military acumen.

To the dismay of Carthage, the Parthians, and Pyrrhus of Epicus, the Roman organizational structure delivered a superior field-level discipline and flexibility—and vastly increased the army's ability to coordinate attacks and deploy defensive measures. Through this structure, the senior leadership deployed the battleground strategy by pushing orders to the field leaders. As the orders arrived in the field, the troops were commanded to achieve specific tactical goals. The army leaders and their subordinates created a rank-and-file order—a model still entrenched in military organizations today. The leader-subordinate relationship is especially important on the battlefield. As leaders fell in the heat of battle, subordinates ascended to fill their shoes—providing on-the-spot replacement and effectively eliminating the chaos normally following the demise of a commander.

The prevailing organizational structure (shown in Figure 6.1) in all types of enterprises is largely based on the military model. Vice presidents and directors are assigned responsibilities for functional or business lines instead of soldiers, but otherwise the structures function remarkably similarly. Executives set the corporate direction and hand control off to middle management for execution. Although the titles may differ, the general structure is the same—a hierarchical pyramid with increasing responsibility and power as one climbs upward.

In Chapter 5, we explored several deficiencies of the contemporary organizational structure, including its outright omission of the customer and its failure to depict how work flows through the organization. To summarize, its real use is to communicate the top-down

FIGURE 6.1 Hierarchical organizational structure.

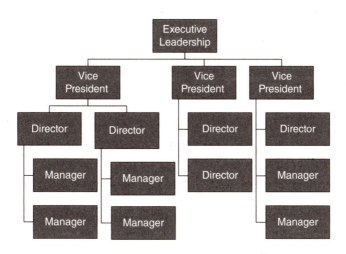

reporting structure from boss to subordinate. But this view does speak to one aspect of modern enterprises—the static and inflexible nature of their human management structures.

Consider the concept of positions in a contemporary enterprise. Once a position is approved and a resource hired, that position is rarely (if ever) reevaluated for adjustment or elimination. In the event of turnover, astute managers immediately generate a requisition to replace the departing employee. They know that eliminating a role translates into a reduced budget and a corresponding reduction in their clout. Because of this belief, managers will fight to maintain their head count in bad times and plead for new roles in good times. Over time, the impact on the enterprise is excessive head count and a workforce misaligned with business needs.

The optimal organizational structure is flexible, responsive to evolving business needs. And further, it reduces ambiguity by clearly defining the roles and responsibilities of performers. This brings us back to the role of process. In a process-focused enterprise, operations are defined by the process system, and the aim of process governance is to align people with the process system.

The concept of process management is foundational to managing a process system. As defined previously, *Process management* is the identification of ownership for every significant process in an enterprise. The individual acting as a process owner is responsible for both the ongoing performance and the innovation of a specific process. What this means is that an individual is accountable for all facets of a process. The process owners oversee the workers who execute the process steps to create outputs. And when a process requires transformation, the process owner participates in the activities to make the desired adjustments. Outside the immediate process team, external resources and other stakeholders may be engaged depending on the extent of the improvements. At the heart of it all, though, is that the primary responsibility for the outputs produced by a process and the ongoing adjustments to that process resides with the process owner. To put it in more granular terms, process owners are charged with designing and developing the process, overseeing execution of the process, and innovating the process from efficiency and strategic standpoints. Luckily, they are empowered by an enterprise built on process.

In most organizations, senior leaders drive any significant change in operational and strategic directions, including how resources are budgeted and the go-to-market strategy. But many leaders are ill equipped to play this role. They are too far removed from the actual work and lack ground-level knowledge of the operational functioning of the enterprise. As a result, leadership directives move about as spotlights across the enterprise—bouncing across departments, markets, and customers often at a frenzied pace as the leadership team responds to perceived threats and opportunities. When bathed in the spotlight, managers diligently attempt to satisfy leadership requests or at least give the appearance of compliance with some of the bizarre directions that drip down from above. Still elsewhere, critically important pieces of the enterprise are ignored and neglected. Where leadership's eyes fall, activity occurs—even when it is horribly misdirected and falls on the wrong areas.

In contrast, an enterprise based on process management receives its direction from the ground up—specifically from the customer. Each of the process owners is an individual point of light—fully versed in the capabilities of their process and equipped to make improvements. Information flows freely from customer feedback points to process owners and strategic teams, where opportunities surface and are debated at the senior leadership level. The resulting initiatives are customer based, consistent with the enterprise's capabilities, and clearly articulated. After the senior leaders prioritize the initiatives and allocate resources, the process owners are integral to completion of the initiatives. Although arguably the most important position in a process-focused enterprise, process owners are but one level of leadership in a process-governance structure.

The process-governance structure (Figure 6.2) consists of five primary roles in addition to the workers who perform the day-to-day activities to keep an enterprise running. The primary differentiator between process and traditional functional organizations is the use of processes as the organizational structure to organize employees—and not titles, functions, or departments. Processes play a role in most functionally organized enterprises, but mostly as the basis for quality improvements and not as an organizational framework.

The layers of the process-governance structure represent five distinct roles or functions that together provide ownership and management of processes. These five layers are the *leadership council* (also called the *process council* or *executive council*), *process sponsors, process owners, process facilitators,* and the *initiative team* (also called a *design team* or *project team*). Each of these roles plays an important part in creating a nimble and innovative organization.

FIGURE 6.2 Process-governance structure.

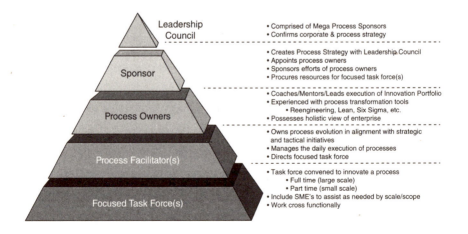

To a large extent, the roles map to the process system, giving life to what is otherwise just a collection of boxes and arrows.

LEADERSHIP COUNCIL (PROCESS COUNCIL)

Similar to any other enterprise, the senior leadership team of a process-focused enterprise plans a market strategy and steers the organization on a course. Indeed, the leadership council plays this role—but in a manner somewhat different from traditional leadership models. The difference is that members earn a seat on the leadership council based on their oversight and management of a megaprocess or they occupy a significant process-based role. In other words, they own and are responsible for a part of the value-creation engine. With very limited exceptions, executive roles without specific and direct responsibility to ground-level work do not exist in a process-focused enterprise. For the leadership council to constructively debate the merits of market opportunities, the collective group must share a detailed understanding of the greater process system in the enterprise and its capabilities. To foster this awareness, leadership council members must possess a ground-level knowledge of the processes in their area—including all the challenges, limitations, and ongoing initiatives affecting their owned area.

Why is it important to connect senior leaders to processes? Well, first off, it allows for informed discussions, as mentioned earlier. Even more important, though, is that it addresses what a former colleague of mine labeled the *ivory tower syndrome*. Inside many corporations, an illness is spreading—and it is ridiculously contagious. It appears in the senior ranks, those leaders who ascended the corporate ladder and hold a position of power and prestige. Having attained such a position, these individuals float in a leadership suspension. When they craft strategies, they do so by hiring consultants or by isolating themselves in a conference room. This seclusion, this hermetic environment in which such leaders exist, is the breeding ground of the next stupid idea. In contrast, when the livelihood of these same leaders is at stake, when they are accountable for a part of the production machinery and its success or failure determines whether they stay employed or get jettisoned to the unemployment ranks, their behavior, especially their level of engagement, changes immediately. In effect, this process-ownership model eliminates the leadership level in corporate America that makes decisions without "getting into the weeds."

Leadership Council Definition

As just articulated, the leadership council leads a process-focused enterprise. Its membership consists of megaprocess sponsors (described in detail later) and a select few senior executives. On occasion, process facilitators (described in detail later) are included as members of the leadership council. Because of their exposure to processes across the enterprise, they offer a unique perspective and an unbiased opinion outside the direct line of process ownership.

Responsibilities

The responsibilities of the leadership council are on par with those scripted to the traditional executive committee. Although variations exist, the responsibilities of the process council include

- Diligently following the customer, market trends, and competitive forces of the industry.
- Fully understanding the organization and its capabilities.
- Managing the competitive position and presentation of the products/services to customers.
- Partnering to build the innovation plan (i.e., the innovation portfolio).

- Serving as the decision-making body to approve or reject ideas/opportunities for evaluation and inclusion in the innovation plan.
- Confirming scope and design of improvement initiatives.
- Setting prioritization determinants to order the execution of improvement initiatives.
- Adjusting the innovation plan to be responsive to marketplace challenges and opportunities in a specific and methodical manner.
- Procuring resources (capital and workers) to execute improvement initiatives.
- Managing the process-governance model by assigning or removing individuals from process-management roles and through the training and coaching of current and future process owners.
- Ensuring that the overall process system is being managed and improved effectively and efficiently.
- Arbitrating issue resolution between process sponsors.

The leadership council, unlike the other roles in the governance structure, is part time. Most leadership councils meet on a monthly or biweekly cadence depending on the degree of change confronting the enterprise. Members of the leadership council spend the lion's share of their time engaged as process sponsors.

PROCESS SPONSOR

The individuals with arguably the greatest ability to enact major change in a process-focused enterprise are the process sponsors. Collectively, these individuals hold the reins to the value-creation engine—guiding the enterprise to a destination while simultaneously ensuring that there is fuel in the tank. They direct and oversee the performance and ongoing improvement of a megaprocess. As managers of a meaningful part of the process system, they are directly connected to the daily activities that drive sales and customer loyalty. In fact, how well they build and manage key processes is a prime determinant of the enterprise's competitive performance.

Process Sponsor Definition

As stated previously, a process sponsor oversees one or more megaprocesses. The role of process sponsor can be sliced into two major areas of responsibilities. First, process sponsors represent their megaprocess on the leadership council and participate fully in the development and performance of the innovation plan. Second, they guide, assist, and coach individual process owners under the umbrella of their megaprocesses. The development of the process-owner level is critical to the flexibility and efficiency of a process- focused enterprise.

The process sponsor's role is similar to that of a business sponsor in most contemporary enterprises. The process sponsor develops and manages direct reports—just as managers in a traditional organization do. However, a process-focused enterprise places a far greater emphasis on process sponsors' knowledge of and engagement with their processes. Fully representing an area on the leadership council requires a thorough understanding of the details of the underlying processes in their area—as well as the activities and limitations of the performers who execute those processes. This knowledge is integral to evaluating and planning improvement initiatives.

Responsibilities

A process sponsor's primary responsibilities include the following:

- Fulfilling his or her responsibilities as a member of the leadership council

- Representing a megaprocess on the leadership council
- Being able to identify the primary impacts that a proposed initiative will have on the processes under his or her leadership
- Educating others on the processes that constitute his or her megaprocess, especially in regard to the interaction between megaprocesses
- Identifying, selecting, and removing process owners in the megaprocess as needed
- Coaching and developing process owners
- Obtaining resources—both capital and workers—to support process owners in their everyday and transformational activities
- Managing issues and cross-process initiatives inside his or her megaprocess
- Working with process owners to identify and achieve efficiency goals

All these responsibilities are geared to allowing an enterprise to capitalize on today's opportunities and to plot a course for the future. Process sponsors set the tone for doing the right things, at the right time, in the right way for their process owners and business partners. Their support for and management of the innovation plan are of preeminent importance in ensuring that the enterprise is strategically and operationally adaptive. The process sponsor brings life to improvement initiatives by allocating the resources to execute initiatives and by coordinating and collaborating with other process sponsors. The connection that process sponsors have with the value-creation process and their responsibility for understanding and reacting to the market put them on the ground floor of transformation.

Identifying Process Owners

The ongoing operations of an innovative enterprise require a healthy pool of candidates with the potential to serve as process owners. As part of the leadership council, process sponsors need to maintain a diverse pool of candidates with the appropriate backgrounds to step into the role of process owner. The aim of diversity in this pool is to support knowledge sharing and collaboration across megaprocesses. Individuals with unique backgrounds and experience approach problems from different vantage points. This helps them to examine the prevailing conventions with a critical eye and eliminates the preponderance of heritage processes in an enterprise. From this pool, process sponsors select the most qualified candidates to be the owners of major processes. Candidates not immediately selected to be process owners may serve as assistants to the process owners (as in *subprocess owners*), or they may be released to participate on initiative teams to further expand their résumé. The key here is an ongoing maintenance of a pool of individuals who possess the capacity and flexibility to fill any number of roles. The importance of the active management of this resource pool is amplified during periods of revolutionary change when new ventures are launched and a process owner needs to hit the ground running.

Procuring Resources

When initiatives receive the blessing of the leadership council, the process sponsor whose area is affected by an initiative is responsible for acquiring the resources to staff the initiative team. To illustrate, if an initiative requires a cross-functional initiative team, the process sponsor works with other affected process sponsors to choose an initiative owner and to source other needed resources. In enterprises that have not adopted the practice of budgeting via improvement initiatives, the process sponsor serves as the budget owner with access to the capital and armed with the political clout to see initiatives through to completion.

Coaching Process Owners

As mentioned earlier, arguably one of the most important contributions of a process sponsor is the mentoring of process owners. Collaboration is of primary importance in a process-focused enterprise. It minimizes the duplication of improvement efforts so prevalent in corporate America and allows process owners and others to leverage the experience of others. The focus of process owners is, as it should be, on the grisly details of their processes. Making collaboration work requires a partnership between process sponsors who focus on the big picture and process owners who are afforded the time to collaborate on improvement activities. With the launch of improvement initiatives, process sponsors monitor progress via an ongoing dialogue with process owners. In this way, process owners benefit from the knowledge and expertise of their process sponsors and ensure that their endeavors are consistent with the grander designs of the leadership council.

Qualifications

In comparison with a functionally based organization, the process sponsor in a process-based enterprise operates on a level similar to a vice president. He or she is a seasoned business professional with numerous innovation initiatives under his or her belt. Although not always possessing the depth of knowledge and experience in all the latest transformation techniques such as Lean, a process sponsor is open minded, humble, and willing to listen and learn. The best process sponsors are enthusiastic, methodical, confident, and stable. They are not looking for quick fixes but rather the *right* fixes. They are patient, understanding, and focused. Although self-confident, they are not overbearing or arrogant. They are fact driven—and logical to a fault. This role is for well-spoken, supportive leaders because a major responsibility is to provide clarity of leadership's intent. For an organization to excel, process sponsors must exhibit the leadership to get involved, become aware, celebrate success, and remedy failure. They must be collaborative leaders, strong communicators, and willing to share knowledge with colleagues and business partners. They are unselfish and results oriented—willing to jettison resources and their personal clout for the greater good. For enterprises that adopt the practices detailed in this book, continual change is inherent. Having a confident captain at the wheel is a necessity.

Executive leaders, usually the leadership council, choose process sponsors. Most process sponsors ideally trace their roots to the ground floor of the enterprise. Throughout their careers, they were in positions to obtain organizational knowledge and build solid business relationships. Their background often includes tenure as process owners or process facilitators before being named process sponsors. Regardless of their path, they have worked in and have an understanding of multiple megaprocesses. A universal perspective is a prerequisite for becoming a process sponsor.

PROCESS OWNER

Process owners are the backbone of a process-focused enterprise. As the owners of processes, they directly manage the value-creation engine. The responsibilities of process owners span both execution and innovation—and, most important, they operate at the level where real change occurs. Unlike traditional organizations, which rarely identify individuals with responsibility for continually improving process performance, a process-focused enterprise places this responsibility squarely on the shoulders of process owners. Their responsibility transcends simply improving the efficiency of a process; the process owner is also responsible for making process adjustments in alignment with the strategic and tactical initiatives identified by the leadership council and communicated via the innovation plan.

Process Owner Definition

A *process owner* is, quite simply, the manager of a specific process. Process ownership may be defined at the major process or the minor process level depending on the scope and the skill sets and knowledge required of the owner. A process owner is responsible for all aspects of a process, including designing the process, obtaining resources (including workers), setting goals, training, and continual improvement.

Even in the smallest companies, there is a benefit to assigning the role of process owner. There may be only a handful of process owners in the most straightforward single-product companies, but the benefit comes from the continual focus on managing and improving the manner in which work is performed. Simply placing the onus for improvement on an individual creates positive momentum and sets an individual up to own ideas on how to improve the process. Further, the named process owner becomes a point of contact for new ideas as well as a knowledgeable resource with whom to discuss improvements.

Responsibilities

The responsibilities of a process owner include the following:

- Owning and acting as the spokesperson for a major process or subprocess
- Developing and capturing customer and worker feedback to build a road map for process improvement
- Understanding the process from the perspectives of customers, workers, and other stakeholders, including awareness of all inputs, outputs, and process steps
- Managing overall process execution
- Monitoring the process's performance via metrics and other performance indicators

- Proactively identifying improvement opportunities
- Being responsible for strategic and operational initiatives that affect the processes under his or her care
- Coordinating and cooperating with other process owners on execution and innovation
- Educating workers and partners thoroughly on all facets of the process he or she owns, including inputs, outputs, processes and subprocesses, competitor processes, metrics, and workers
- Identifying (frequently with the assistance of a process sponsor), selecting, and coaching process and subprocess owners and, when circumstances dictate, removing individuals from their positions
- Training and developing workers on process knowledge and skills
- On large-scale initiatives, partnering with process facilitators and initiative teams to drive improvement in the process and in alignment with other enterprise initiatives

Although process execution, design, and alignment are the primary responsibilities of process owners, process owners are also responsible for ensuring that these activities do not occur in a vacuum. Insular leadership is dangerous to a continually evolving enterprise and is a key reason why functional organizations react so poorly. Aligning and optimizing their processes with interconnecting and interdependent processes is a key responsibility of process owners. When change agents fail to take the big picture into consideration, an enterprise runs the risk of optimizing a piece of the system at the expense of the whole system. The risk of localized improvements is further mitigated through the involvement of process facilitators (discussed in the next section) to lead cross-process initiatives and by setting metrics to measure the full end-to-end performance of the process.

Depending on the scope of the change, process owners may complete smaller, bounded initiatives on their own or with the support of a process facilitator. For large-scale transformation efforts, going at it alone is unrealistic because of the time commitments of the process owner and the need for objectivity (i.e., a process owner's vantage point may be distorted by his or her ownership). Most large-scale transformational efforts are cross-functional in scope—spanning multiple departments, business units, and perhaps geographies. In order to generate the most beneficial results, a cross-functional initiative team of knowledgeable and capable individuals may be commissioned to analyze, design, develop, and implement major transformations. When the initiative-team approach is employed, process owners still play a key role in the initiative—but as customers of the initiative team and as subject-matter experts when their knowledge is required.

Qualifications

Process owners operate at a level largely on par with a departmental or business-line owner in a traditional organization. But there are significant differences in their qualifications and their training prior to being selected to serve as a process owner. A functional leader may be promoted to a position without the training and experience required for the role. This quirk of the corporate world inspired the adage, "It's not what you know, but whom you know." In contrast, the path to becoming a process owner requires an individual to earn the role through demonstration of the appropriate skills and knowledge over a period of time. Generally, individuals progress through varied process ownership roles with a gradual escalation in size, scope, importance, and difficulty of the owned process. Promotions take time. Individuals must earn their strips in the trenches—earning increased responsibility by demonstrating aptitude in getting results.

The path to becoming a process owner weaves around curves and offers a multitude of branches from which to choose. An individual might begin their path as a member of an initiative team—working on a strategic initiative to launch a new business line. Through this experience, the individual gains a ground-level understanding of processes and how they connect and build on each other to create utility. On the project's completion, this individual may well be a candidate to work on the initiative's subject processes as a subprocess owner. And the progression continues. After time in the process-based role, this same individual may transition to roles on tangential processes or subprocesses or may even bounce back to work on another initiative team to further grow his or her knowledge and skills. Over time, the individual gains not only process-management and process-improvement skills but also an awareness of the overall process system and the interconnectivity of points throughout the enterprise.

As mentioned previously, process sponsors appoint process owners—and there are often a number of vacant positions. This is because turnover in the process-owner ranks is not only expected but also encouraged within reason. By rotating process owners in and out of positions, process sponsors ensure that the underlying processes periodically receive the benefit of a fresh pair of eyes. As a result, the enterprise enjoys continual reevaluation of major processes and a cross-pollination of ideas while also benefiting from deeper cross-functional linkages. This is not to say that the role of process owner should be a revolving door. Optimally, process owners remain in their role for a minimum of two years—although the duration differs based on the size and scope of the specific process. After this first assignment as a process owner, an individual may slide across the organization to become a process owner in a related process or perhaps become a process facilitator to further hone his or her transformation acumen. When a process owner becomes sedentary in a role and begins to resist innovation, it is time for him or her to move to a new assignment. Naysayers are not good process owners. The intention of the role is to eliminate and not create organizational silos. Turnover helps to reduce "groupthink" and creates opportunities for the next generation of leaders. It propagates institutional knowledge, increases organizational fluidity, and reduces the number of employees who leave the company by providing challenging roles.

Process owners are the stars of a process-focused enterprise and are critical to its performance. They possess a passion for perfection and willingly dig deep and work late to get answers. They are risk takers—always seeking the next great thing. As seasoned performers, they possess strong business acumen, but they are far from rigid in their beliefs or actions. They are out-of-the-box thinkers who are not scared of change—but delighted by its prospects. They are leaders and collaborators who have the ability to switch roles when the situation requires it. They are enthusiastic lifelong learners and willing to reconsider their perspective. They are engaged, methodical, structured, and motivated.

PROCESS FACILITATOR

Process owners are responsible for keeping their processes humming and ensuring that the proverbial lights stay on. Additionally, they hold the reins for completing small-scale initiatives—mostly those confined to the boundaries of their processes and focused on reducing cost, improving quality, or enhancing scalability. Whereas a process owner may well benefit from coaching, he or she is generally able to tackle localized efforts without significant assistance.

On occasion, though, the winds of change escalate to hurricane levels—pushing leadership teams to react fast and big. Small, measured steps are inadequate to staying alive. Under such conditions, process owners are insufficiently armed and resourced to perform their real job while simultaneously spearheading major transformation efforts. A special forces unit is required—a team led by a battle-tested expert. Such an individual in a process-focused enterprise is a *process facilitator*.

Process Facilitator Definition

A process facilitator is the catalyst behind the execution of game-changing (or even simply survival) initiatives. Process facilitators are the battle-scarred veterans who have seemingly seen it all. Their domain is change—and most often on a big scale. They come armed with the background to complete big transformational efforts such as the launch of a new business unit, the overhaul of an existing business unit, or a foray into a new market. They may not be content experts, but they have the institutional know-how to steer major endeavors. Their roles are numerous—leader, coach, process expert, teacher, mentor, subject-matter expert, black belt, Lean expert, and Reengineering practitioner, among many others. They are well versed in the phases of solution creation—masters at analyzing, designing, developing, and implementing solutions. They are worldly and able to add value in any environment. To put it succinctly, they are the right person to tackle the biggest challenges. They are the lubricant of innovation.

A process facilitator's degree of involvement with an initiative expands with its scope and complexity. As mentioned earlier, for small-scale improvement efforts, a process facilitator may coach a process owner on how to approach the initiative, or he or she might offer an outsider's perspective on how to improve a process. For the most part, though, process facilitators keep their hands clean when it comes to smaller initiatives.

Where process facilitators typically engage is on the stickier programs—those larger, more daunting initiatives that require more than just a part-time coach. These improvement programs span multiple functional areas and involve a handful of process owners and subject-matter experts. As the size, complexity, stakeholders, and risk increase, the expertise provided by a process facilitator becomes increasingly important.

For such initiatives, the process facilitator may be engaged to coach the team through the activities required to analyze, design, pilot, adjust, and launch the solution. On occasion, a process facilitator may lead a major initiative, but this occurs only in extreme cases—mostly when an enterprise is in survival mode. The true intent of this role is to serve as the glue between major initiatives, ensuring that the teams collaborate to build holistic solutions that address all the intricacies of the enterprise. Having process facilitators lead major initiatives gradually weakens the institutional breadth of their knowledge and reduces their effectiveness in designing cross-functional solutions. Ideally, they work on several big initiatives at once—sharing their experience and awareness of the changes occurring across the organization to ensure connectivity, reduce duplication, and allow for the prompt identification of synergies. In this way, the role of process facilitator provides an objective and informed perspective for an enterprise's innovation capabilities and ongoing improvement activities.

From this facet of a process facilitator's role follows perhaps its greatest contribution to a growing enterprise. Leaders may well understand the opportunities available to an enterprise—including both those of a strategic or tactical slant. However, their knowledge of their enterprise's capabilities at the ground level is frequently limited. The result is a knowledge gap—and occasionally a huge chasm—between strategic intent and everyday execution. And this gap takes the form of unclear strategic directives.

The unclear strategic directive is the most pervasive and troublesome challenge for the eventual owner of any improvement initiative. Leaders expect results but commonly provide little or no direction as to how to achieve them. The gap between *direction* and *results* is vast today—leading to lackluster strategic performance across corporate America. Most contemporary enterprises lack an individual or team that is able work across processes, departments, and organizations to move the enterprise beyond loosely worded ideas to actual results. A bridge is imperative to connect great ideas with great execution.

Into this void steps the process facilitator, an individual with the knowledge and the vantage point to take leadership intentions and translate them into an actionable plan. Process facilitators are uniquely positioned in the enterprise to understand its operational capabilities but sufficiently distanced from the daily rigors as to not be persuaded by politics, current mishaps, or other short-time issues. Not only do they understand the operational component of change, but they also see firsthand the daily impact of external forces threatening the enterprise. The advantages of having a process facilitator immediately become apparent when an enterprise embarks on a large-scale strategic initiative. To respond appropriately to an opportunity, someone needs to design the end solution, including assessing how it affects all the component parts of the enterprise. The design's benefits, as well as all the development and ongoing costs, need to be folded into a business case. And from this business case will come the final determination as to whether the project should move forward or be discarded. In most contemporary enterprises, there is no one to shepherd the development of large-scale initiatives and ensure that the business case truly represents a thorough design, all the downstream impacts, and the corresponding costs. Process facilitators working in tandem with process owners are a powerful combination to tackle this assignment and give a degree of confirmation that enterprise resources and energy are focused on worthwhile endeavors.

Responsibilities

The responsibilities of a process facilitator include the following:

- Attaining and maintaining expertise-in-transformation tools and methodologies—especially tools aligned with the upcoming needs of the enterprise
- Understanding the greater organization from the perspective of customers, workers, and other stakeholders, including awareness of all inputs, outputs, and major processes
- Acting objectively in the interests of the enterprise and facilitating discussions, debates, and decisions on large-scale improvement efforts using a fact-based approach
- Coaching process owners through the delivery of limited-scope efficiency efforts
- Supporting large cross-functional transformational efforts by actively coaching or leading teams through the execution of strategic or operational initiatives

Qualifications

To be successful with their responsibilities, process facilitators need a holistic view of the organization's operations as well as a deep understanding of how strategic and tactical initiatives are executed. Understanding the interactions between processes and possessing a thorough understanding of process-transformation tools allows them to design and implement solutions that improve the process system.

Process facilitators thrive in chaos, can pull order from disorder, have a can-do attitude, are focused and passionate, and are ready to roll up their sleeves and get dirty. They are leaders, drivers, coaches, motivators, and persuaders. They work smarter and harder—earning the respect of both process owners at the ground level and senior leadership at the upper levels. They readily admit mistakes and then quickly transform into problem-solving mode. They praise the successes of their business partners and take the blame for the failures. In short, they are prepared and motivated to drive initiatives toward success regardless of the challenges they face or the acclaim they receive.

The background of process facilitators varies but almost always includes experience in a consulting role. Although they may not be a recognized expert in process transformation, they are highly experienced and have completed projects using one or more of the major transformational tools such as Six Sigma, Lean, Total Quality Management, and Reengineering, as well as value-chain analysis, organizational design, and change management. They are often worldly in their work experiences—having completed tours of duty across organizations, industries, and geographic boundaries.

Although their background may mirror that of Six Sigma black belts at companies such as Motorola, General Electric, and Bank of America, they are more broadly based in their experiences and rarely focus on only a single initiative at a time. And although most black belts focus primarily on the Six Sigma toolset, a process facilitator uses an immensely larger toolbox. Process facilitators are aware of the intricacies of different tools and their appropriate usage.

There is no rigid career path for process facilitators. Often their personalities lend to bouncing about organizations. Although they may enjoy the work and lodge permanently in a process facilitator role, they might transition into process-owner or process-sponsor roles, especially when the new position is tasked with audacious goals. These are individuals who want to be on the front lines in the heat of the battle. As a result, they are often associated with initiative teams.

Initiative Team

Initiative teams are the special forces of the business world. When projects and initiatives cross departments/business units, are complex, require specialized knowledge/skill sets, or demand immediate results, an initiative team is frequently the answer.

Initiative Team Definition

An initiative team is the heavy-lifting group in a process-focused organization. This team goes by a host of names, including *transformation team*, *project team*, *design team*, *special-project team*, and *task force*. For the most part, all these titles are accurate and appropriately reflect the potential roles this group fills. However, adding *short-term* to the title may be a touch inappropriate. An initiative team's work is to develop a solution that generates the greatest value given the circumstances. Only on rare occasions should the goal be to provide a short-term fix because such fixes rarely translate into long-term advantages.

Initiative teams are appropriate for initiatives that are cross-functional, require a high degree of coordination and collaboration, are entrepreneurial or a new venture, or are highly complicated. In short, an initiative-team approach is appropriate when an initiative has such a scope or complexity that a single process owner or small group of process owners do not have the time, knowledge, skill sets, objectivity, or authority to drive it home.

Be prudent in the use of initiative teams. Avoid the common mistake of launching initiative teams under the premise that some activity is better than none. Any action without a real destination is just another form of waste. Initiative teams should never be formed if the work does not warrant it. Many initiatives can be addressed with only a handful of engaged individuals—usually a handful of process owners and a process facilitator. These efforts do not constitute an initiative team because the complexity and cross-functional requirements are minimal.

There is an exception: initiative teams can be effective mechanisms to complete initiatives when politics (both internal and external) so muddy the water that even the smallest iota of progress is bought with an inordinate amount of sweat and tears. And this is not an uncommon event in corporate America's offices. Leaders ardently protect their sliver of the organization and aggressively assert their perspectives—often sowing confusion and thwarting progress. The results are costly from both financial and strategic perspectives. Because initiative teams do not report directly to any single leader but rather to the full leadership council, the impact of political turf wars is greatly lessened. The intent is for initiative teams to operate independently—to act objectively and do the right thing for the enterprise rather than comply with the direction of any single leader.

Once an initiative team is commissioned, the process sponsor, with the support of the leadership council, is responsible for staffing the team. Ideally, a sponsor matches content and aim of the initiative with individuals equipped to be successful. Still, additional considerations may factor into the selection of team members. For example, with a cross-functional scope, it is important to include team members on either the core team or the expanded team with ground-level awareness of the specific content area of the initiative's scope. Diversity is another consideration because research shows that teams with diverse backgrounds, perspectives, and opinions produce superior results—as well as mitigate the risk of "groupthink." Research also suggests that team members with prior experience working together often jump-start the effort, allowing it to get up and running much faster than teams that must first establish familiarity and trust.

An initiative team may be commissioned full time for a major initiative or part time for a smaller initiative with minimal complexity. Look to staff the core team with five to seven members for average projects. When the size of the team expands beyond seven, communication, coordination, and decision making become increasingly challenging—any gains in team productivity from having more hands is offset by the increased difficulty of organizing, coordinating, educating, and aligning the team.

When building an initiative team, consider the scope and complexity of the initiative. Identify the requisite knowledge, skill sets, and experience that would be beneficial. Leaders commonly handpick team members from every conceivable stakeholder group (regardless of their relevance to the initiative's scope) or, even worse, never really organize resources around the initiative. Equally likely, team members are selected based primarily on their current availability. Be judicious. Get the right resources and not the available resources.

A final consideration when staffing initiatives is whether to use external professionals. External professionals are more than just consultants. They may be full- or part-time team members from outside the company, including resources from academia (who may join the team during sabbaticals or during summer vacation), industry experts, or representatives from business partners. Introducing a few external professionals into the mix is a good practice. Not only might they provide skill sets and knowledge not available internally, but they also bring a perspective that is untainted by the enterprise's internal norms and practices.

Even after including external professionals, many initiative teams still lack specific knowledge or understanding of the subject matter. For this reason, initiative teams may (and in most cases should) also include an extended team of subject-matter experts who engage on a part-time basis. The role of these individuals is to provide specific knowledge or insight on components of the work product. In most cases, they do not need to be full-time team members but rather float in and out as their input is needed. Their participation may encompass only a single design session. However, as a solution's design takes form, the initiative's scope might shift, requiring a part-time resource to engage on a full-time team basis. Flexibility is the key to ensuring that the right mix of resources is available to deliver the solution. It falls under the responsibilities of the leadership council to develop a resource pool that is scalable and adaptable to the continual innovation needs of the enterprise.

Responsibilities

When initiatives or projects require a specialized team, a focused initiative team is organized and set loose to bring it home. On commissioning the team, the responsibility for innovation is transferred from a process owner (or multiple process owners) to the initiative team. From this time until the initiative is complete, the initiative team acts as the de facto process owner, with the exception of the daily execution of the subject process. This means that the initiative team is tasked with the role of representing the process(es) in innovation discussions and collaboration events with other initiative teams or process owners.

When the initiative team convenes for the first time, the sponsor is responsible for sharing the rough scope of the initiative and all the available details. At this point, the team takes over and is responsible for analyzing the current situation, developing a solution, testing and adjusting the design, piloting the design, expanding the pilot into the launch of the solution, and finally transferring control of the fully developed solution back to the process owners. The transference back to the process owners includes all aspects of the solution that, in their entirety, allow for the ongoing support and execution of the process(es).

Qualifications

Process sponsors select initiative-team members with the goal of aligning their capabilities with the initiative's needs. One of the most immediate challenges is to select team members who are knowledgeable about key processes or functions included in the initiative but who are also open minded and willing to put aside any sense of ownership of the existing state. To achieve this arrangement—particularly for organizations that have only partially implemented a process-focused organization methodology—process sponsors should avoid selecting process owners and major stakeholders from the processes and functions that fall within the initiative's scope. However, it is imperative to include individuals with a solid worker-level view of the processes undergoing analysis. My recommendation is to select team members who have worked on processes that are tangential or closely connected to the processes under investigation. A prospective team member's prior role may have been with a process that provided an input or received an output from the subject process. In this way, the team possesses knowledge of the process but retains the objectivity required to break down barriers and develop innovative solutions from a blank slate. Another approach that is often used in tandem or in place of this approach is to include the process owners and subprocess owners on the extended team—bringing them into the loop when their subject-matter knowledge is helpful.

Considerations for selecting team members include not only their abilities and knowledge but also such factors as diversity, individual relationships, and personal characteristics. The most successful initiative team includes members with personal attributes such as the following:

- Fact based
- Out-of-the-box thinkers
- Dedicated to continual achievement
- Challengers of the status quo
- Overworked and overstaffed because of their knowledge, experiences, and capabilities
- Good listeners
- Synthesizers of differing opinions and approaches
- Connectors/networkers

Because of the nature of the work, an initiative team assignment is transitory. The rationale for launching an initiative may crumble as current events force the enterprise to move in alternative directions.

Every initiative is always at risk of being delayed or even eliminated. Individuals assigned to initiative teams sign up with an acceptance of this risk. Therefore, the role of an initiative-team member is rarely a long-term commitment. Individual team members may play a short-term role, staffed on initiative teams based on their ability to fulfill a need and released on completion to pursue other opportunities.

The evolving nature of the work requires a pool of people with talents befitting the enterprise's cumulative innovation needs. Although life in this resource pool comes with challenges, it also offers benefits that rarely exist in a more traditional position. The most prominent benefit for the individual is the ability to act as an internal consultant—working on varied projects and honing valuable skills. But the enterprise benefits as well. Individuals working on major initiatives form relationships with individuals from other corners of the enterprise. At the same time, initiative-team members gain an education on how distinct processes fit together. The relationships developed and the acquired knowledge are undoubtedly the best training an individual can obtain to position himself or herself for a myriad of roles. For this reason, initiative teams inadvertently function as a training ground for future leaders. Simply by looking at the qualifications to be a process owner, it is easy to see why members of initiative teams routinely find themselves asked to transition into process leadership roles.

As a process-focused enterprise matures, a good number of the individuals in the pool of resources for initiative teams continue to remain in the pool because of their personalities and their love of project work. For these individuals, continued learning and evolution of their skill sets are paramount to their happiness. They just are not cut out for line work. Still others float through initiative teams because no position is currently available that aligns with their backgrounds. Some organizations even elect to hire new resources into rotational initiative teams as a way to assimilate them into the enterprise.

Ideally, an enterprise opts for a healthy mix of full-time members of initiative team and a set of individuals rotating through for the learning experience or until they find another role. As with any other organization, initiative teams can get stale without a continued stream of new talent. When the pool is managed well, it becomes an engine of collaboration and a trove of institutional knowledge. The size and depth of the pool's collective individual talents are directly correlated with an enterprise's innovation capacity.

HUMAN RESOURCES MANAGEMENT AND THE PROCESS-GOVERNANCE MODEL

In comparison with the traditional functionally based organization, a process-based structure requires a slightly different approach to human resources management. The traditional organization has a rigid hierarchy that allows for a clear delineation of supervision. In a process-focused enterprise, where individuals are assigned to processes, individuals now report up through the process-governance structure.

The prevailing structures and methods practiced in most contemporary human resources departments are continuously questioned as to their effectiveness. Often the judgments are harsh. Critics point to a number of routine practices that are believed to demoralize employees and increase turnover. Job placement is frequently cited to be more about whom a candidate knows than what he or she knows. Compensation and reward systems lack connectivity to employee performance and value creation and, even worse, sharply penalize employees for taking risks. Performance-appraisal biases are widespread, and workers spend more energy pleasing their bosses than creating value. Training programs are panned because their content is generic and not relevant to employee roles. Individuals are routinely promoted into supervisory roles without prior management training or experience. Overall, the average employee is disengaged, frustrated, and often looking for positions elsewhere. To many, anything would be a step up from their current situation.

With a process-governance structure, a tremendous opportunity exists to correct existing deficiencies and reengineer human resources practices for even greater effectiveness in the future. The management pyramid with clear lines of supervision is replaced with a process-based structure. But the structure is constantly adjusted based on business priorities. Individuals move in and out of roles based on business needs. Management structures are fluid and flexible. Without the hierarchical structure, the clear path up the corporate ladder no longer exists. In its place is a structure in flux, where roles exist for several months and then potentially disappear in step with market shifts. And while the fluidity of the model complicates traditional managerial oversight, it opens up a new realm of opportunities to staff, train, and motivate employees.

MANAGEMENT

Although a loose structure of management exists in a process-based organization through the process-governance structure (e.g., process sponsors, process owners, and other associates assigned to processes), the organization's continual strategic reinvention and resulting movement of employees often result in only temporary reporting relationships. A new style of management is required. The key theme of management in a process-based structure is that feedback and coaching to employees need to come from not only an immediate supervisor but also others in contact with the employee. To this end, the value of 360-degree feedback is multiplied in a process-based environment simply because of the fluidity of the organizational structure.

The primary guidance for all employees in a process-based organization is the innovation plan. This plan prioritizes the initiatives to direct the enterprise. Because the initiatives are based on process output attributes, it is easy to communication innovation goals to teams for execution. The importance of teams in a process-focused enterprise necessitates training on team building. A critical component of building highly performing teams is straightforward and open communication—providing feedback and coaching promptly and empirically. This is one of the primary avenues by which individuals receive feedback in a process-focused environment.

After working in a process based environment, I discovered one of the greatest opportunities for individuals to grow in their roles and assignments is through cross-process role sharing. An enterprise consists of a considerable number of process owners. Although they work in different areas, there are significant commonalities in their role, and they benefit from shared learning experiences. For example, understanding and improving process execution are important to every process owner. And the human side of work, although always unique, includes the same basics of motivating, coaching, and working with individuals. To share experiences across processes, individuals at a level (e.g., process owners and subprocess owners) ideally meet on a regular basis to share their successes, failures, and as well as the opportunities confronting them. When needed, individuals can request the counsel of their colleagues—using the group's collective experiences to brainstorm alternative courses of action. This type of assembly is of tremendous benefit to individuals in their quest for personal and corporate growth. Although possible and beneficial in a traditional organization, this type of assembly provides even greater benefit in a process-focused enterprise because it forges a connection between individuals in roles—allowing for knowledge sharing and collective improvement.

CAREER PATHS

The performance and viability of the traditional career path are largely predicated on the quality of managers today. If managers take an active interest in developing the employees in their reporting structure and groom them for roles both internal and external to their business units, the hierarchical structure can work quite well at growing a workforce. However, anyone who has spent time in the business world identifies the fallacy of this logic right off the bat. Very few managers take it on themselves to groom their employees for other roles. This is not because they are not an altruistic bunch, but rather the demands of running a business direct their attention elsewhere—to the burning platforms. With the exception of when the development of an employee dovetails with a dire corporate need, an individual's growth is largely driven by his or her motivation.

Most individuals at a corporation begin their career on the lower rungs of the corporate ladder. The mantra is that they need to learn the ropes to ascend further. The pace and duration of an employee's climb are supposedly tied to their accumulated knowledge and experiences. However, in reality, this is seldom the truth. A sizable number of employees find themselves plugged into a functional pigeonhole—exposed to a narrow slice of the enterprise—often defined by a functional boundary. The individual's career path and available opportunities are largely influenced by the perceptions and biases of a single immediate supervisor. An outgrowth of this system is a collection of work rules that are fairly standard across enterprises, including

- A rigid hierarchy exists, with employees set in roles with limited authority but held accountable and rewarded based on enterprise results.

- Risk taking is dangerous. A failure can doom a career, whereas success can be quickly minimized or claimed by others.
- Complacency is the norm because organizational forces discourage bold or significant action.
- Individuals are ill prepared to cope with change and have limited opportunities to expand their skill sets in process techniques or other improvement methodologies.
- Work networks and relationships are developed based largely on a worker's motivation and not a part of enterprise practices to encourage cross-functional knowledge sharing.
- Career prospects are compartmentalized to functional areas, and many individuals with lofty aspirations are forced to look externally for personal growth opportunities.

A process-focused enterprise is radically different in its approach to managing, developing, and using human resources. In the absence of a corporate hierarchy, individuals are faced with a somewhat ambiguous career path. Roles are drastically different —no longer identified by department and position title but now linked to processes. And roles come and go based on business needs. In a process-based environment, employees are described by the roles they play (e.g., initiative-team member, process owner, process sponsor, line worker, or customer-service representative) and the capabilities they possess (e.g., statistical process control or new-product development). Matching resources to enterprise needs requires a thorough understanding of resource capabilities and initiative requirements. When both are known, making a match is fairly simple. From an employee's perspective, this system requires a greater degree of self-management. Similar to choosing college classes to meet graduation requirements, employees are given the opportunity to manage their careers by seeking opportunities and acquiring skills that prepare them for future opportunities. To illustrate, let us consider two examples of individuals managing their careers in a process-focused enterprise.

June is an entry-level analyst. Her career began in a role supporting the accounts-payable process. Two years into that role, she was provided with an opportunity to become a subprocess owner—specifically focused on the mailroom processes. June worked hard, leveraged a Lean expert to broaden her skill base, and was successful at cutting costs in the mailroom by 20 percent. Based on her success, she was asked to become a member of an initiative team focused on expanding the company's shared service model. June enjoyed the work and further expanded her capabilities. But her desire was to return to accounts payable as a process owner. A year later, the process-owner role became available, June applied for the position, and—based on her history of success with the process and her experience with innovation methodologies—she was hired.

Corwin is an experienced business manager. He was originally hired to oversee the launch of a new business line. Corwin had decent process skills, but after six months in the role, he came to the realization that he would benefit from further process training. Shortly thereafter, Corwin was assigned to an initiative team executing a reengineering effort on an underperforming business function. For the next year, he was heads down—learning about process transformation and rebuilding the function in line with market opportunities. At the end of this tour of duty, Corwin found that he enjoyed the fast-paced environment of initiative teams. He began training to become a process facilitator—a role he ascended to after three years. Eight satisfying years later, he opted to return to the front lines and accepted a position as a process owner. Ironically, it was the same process he had opted out of when he joined the company—although since that time the business line, had matured, and planning was well underway to expand into international markets.

June and Corwin illustrate two potential career paths in process-focused organizations. Not only did their roles vary as they progressed along their career paths, but they were eligible to move into roles across the full organization. The career paths illustrated in both examples allowed the employees to gain process-improvement skills and be exposed to multiple areas of the enterprise. This model benefited not only the employees but also the enterprise because it got to reap the rewards of the employees' development.

A human resources goal in a process-focused enterprise is for roles to be flexible enough to meet business needs but also to be relatively consistent across levels. For example, a process owner shares similar responsibilities to a process owner lodged elsewhere. Of course, the content of their respective processes may be radically different. The benefit of role consistency is that it yields workforce flexibility and minimizes the time it takes for a process owner to get acclimated in a new role. And just as in any other structure, some individuals—especially those with a depth of experience in one functional area—may have no desire to leave their current role—and it may be equally beneficial to the enterprise. However, such functional experts are likely to be solicited for participation on initiative teams to deepen their organizational awareness and allow the enterprise to capitalize on their expertise.

One constant risk for all enterprises is that individuals who are long in a position may become blind to improvement opportunities. This risk is significantly reduced in a process-focused enterprise because of the fluid nature of career paths, and it can be further mitigated by the periodic rotation of skilled individuals to initiative teams. Pulling employees away from their regular work in essence provides them with a sabbatical and gives them an opportunity to enrich their knowledge and abilities—ideally to benefit their regular role. Once employees get used to this model and the flexibility it provides, job satisfaction and employee morale increase because of the prevalence of personal development opportunities through new roles and through continual learning and skill-set enhancement.

COMPENSATION

The flexible nature of the workforce in a process-focused enterprise complicates the existence of a traditional compensation policy. In today's corporate world, employees largely negotiate their compensation when they accept a position. Any increases thereafter are based on corporate or individual performance—largely driven by the perception of an immediate supervisor. With continually evolving roles and a reduction of supervisory oversight, supervisors are even less able to accurately align an individual's rewards with his or her efforts.

A process-focused enterprise accommodates the use of a human resources management program that more closely aligns an employee's performance with his or her compensation. One of the major aims of a process-based model is to organize employees around value-creation activities and thereby significantly reduce wasted investments in cost, energy, and resources. To the benefit of the employee as well as the enterprise, recognizing value creation at a process level allows for the capture of a significant amount of empirical data that is useful for understanding the performance of a team or an individual. Such data is invaluable in creating an objective and straightforward compensation program.

In most positions, the lion's share of a compensation package is the base salary. In contemporary large enterprises, salary bands are often created to simplify and introduce fairness into the compensation structure. Every position is graded into a band based on its responsibilities. Once the range is set, hiring managers can negotiate a position's salary with applicants within this predetermined range.

This model translates well into process terms. For example, process owners might be placed into one of three bands. The bands may be segmented depending on the scope, risk, complexity, or skill sets necessary to oversee the underlying process. Process owners in the first category oversee processes with multiple subprocess owners, have substantial risk and complexity to the process, and are core to the enterprise's success. Process owners in the second category might manage processes with a good amount of risk and, although important, are not comparable with the process-ownership roles in the first category in terms of importance to the overall enterprise. The final level of process ownership includes process owners with processes that require ongoing management but are not of significant risk or complexity to merit more than a junior process owner. Subprocess owners might also fall into this third group depending on their responsibilities.

Human resources professionals are critical to setting up the bands and their corresponding base salaries and monitoring the bands on an ongoing basis to determine when salary or role adjustments are appropriate. In addition, human resources professionals monitor the overall labor market and make recommendations when universal adjustments such as cost-of-living increases should be incorporated.

If the base salary is the compensation an individual receives for performing an enterprise-required role, *variable compensation* is the pay an individual receives for doing more than is expected and doing it better. Variable compensation is the manner by which high performers are differentiated from the remainder of the workforce. The key to an effective and fair variable compensation program is accurate information on the performance of an individual relative to his or her peers and the tasks the individual was assigned. Fortunately, process-based goals enable greater measurability and tracking than any similar performance-monitoring mechanism.

In a process-based environment, the variable compensation component can be tied to process or initiative performance, allowing an individual's efforts to be tied to an outcome. To have integrity, process and initiative goals need to be clearly stated on the process action to be completed—and not necessarily the outcome. For example, a process outcome is to adjust a process to produce a new offering, or it may be to identify ways to reduce the cost of the process. Sample process outcomes based on their intent include the following:

- *Process (where the goal is strategic)*. Actual adjustment of the process or an increase in sales or market share.
- *Process (where the aim is to improve efficiency)*. Cost, quality, throughput improvements.
- *Process (where the intent is to expand existing capabilities to compete better in the future)*. Development of process scalability, flexibility, and adaptability or simply the creation of a new process.

When the process goals are accomplished, the involved individuals receive a predetermined reward as a bonus. In the same way, when initiative teams fulfill their objectives, they may receive a bonus for the incremental value added by their efforts. When identifying the bonus for a process outcome, the risk, scope, and complexity of the process should be incorporated into the calculation.

In many instances, the benefit of a process improvement or initiative may not be realized immediately—for example, the intent of a process outcome may be to expand the adaptability of a process to support more customized outputs. In such instances, a compensation specialist may need to be engaged to ensure that the bonus awarded on successful completion of the process adjustment is appropriate.

LEARNING AND DEVELOPMENT

With any process-management undertaking, one of the most pressing needs is to get the full workforce educated on basic process terminology and improvement toolkits. Unfortunately, process-improvement capabilities are vastly undeveloped in most companies. When the economic road is bumpy, training is a frequent victim of cost-cutting initiatives. This deficiency has an impact on the design and management of enterprise processes. Major processes in many companies were never designed for the functions they perform. They are heritage processes—lacking a formalized or methodical design. Education is paramount to eliminating negligence in designing and managing processes. The aim of process training is to create a widespread base level of process know-how. Initial process training includes

- Process terminology, including the basics of the process-focused organization.
- Introductory process skill sets, including flowcharting, process analysis, informational interviewing, and process design, testing, piloting, adjusting, and launching.
- Financial modeling to build business cases.
- Basic change-management training.
- Deployment and implementation training.

As individuals become comfortable with the basics, more advanced curricula provide skill enrichment for experienced process owners, process facilitators, initiative-team members, and others. An advanced process curriculum focuses on skill sets, including

- Customer analytics and connections.
- Process design and transformation.

- Improvement methodologies, including Lean, Six Sigma, Agile, and others.
- Organizational design.
- Change management.

Ideally enterprises design their learning curricula using employee roles as the basis for training invitations. This ensures that the content is geared to individuals needing the knowledge for their current or future role. Additionally, many enterprises offer training programs that cover the specifics of the enterprise's organization. This training might include the enterprise process blueprint, the process-governance organization, and the innovation plan in a process-focused enterprise.

All that aside, learning and development do not need to be limited to the classroom setting. The process-focused enterprise intrinsically provides a number of options for skill and knowledge enrichment, including on-the-job training, cross-process role sharing, coaching and mentoring relationships, and the process organization itself (e.g., serving as a subprocess owner or on an initiative team).

ANTICIPATED BENEFITS OF A PROCESS-GOVERNANCE STRUCTURE

The creation and implementation of a process-governance structure often give rise to a number of immediate benefits because roles are clarified and the enterprise's innovative capacities receive their just treatment. The anticipated benefits of a process-governance structure include the following:

- *It provides clarity of ownership.* An immediate benefit of a process-ownership model is the clear identification of the owner of a specific process. As employees uncover information or generate ideas regarding that process, they now know who to ferry this information to in order to get it addressed.

- *It fosters collaboration.* The process-governance structure, by its delineation of ownership alone, facilitates collaboration and knowledge sharing across roles. Additionally, the aim of the process-governance structure is to place the right employees with the right skill sets where they will generate the greatest value. This aim in and of itself encourages employees to move from role to role, expanding their institutional knowledge across the enterprise and aiding informed decision making at all levels of the enterprise.
- *It promotes an engaged and motivated workforce.* The structure drives responsibility, empowerment, and accountability down to the ground level and mobilizes the troops to a unified enterprise wide improvement agenda.

Although numerous benefits result from the increased clarity and understanding of roles, one of the major downstream impacts of a process-based approach is the ability to implement an objective performance-management system—arguable one that is most akin to a true meritocracy, where individuals are appropriately rewarded based on their individual contributions to the enterprise's fortunes.

For a moment, think about the traditional organizational setting, where an employee works in a functional department such as finance. When the boss moves on, a replacement is sought from the available qualified resources. There are instances when the exact right individual is promoted into the role, but more often than not a better candidate exists. Why is this? My thought—because of a scarcity of empirical data to really evaluate individual performance, promotions (and rewards) are largely based on subjective factors. In effect, today's corporate advancement is effectively a popularity contest.

In contrast, a process-focused enterprise has the data to evaluate individuals based on their readiness for a role, as determined by their prior roles and performance. The entire process system and promotional process are predicated on individuals moving to expanded roles, performing in those roles, and then being promoted to roles of increased responsibility. In this way, leaders are not placed in roles for which they are not adequately prepared. No organizational structure and performance-management system arms individuals to ascend to new roles and challenges as well as a process-based organizational structure.

The process system and process-governance model are the organizational structures of a process-focused enterprise and are imperative to its smooth functioning. If competitors never sought to nab market share, if customer preferences were stagnant, if raw materials prices never moved, these structures would be sufficient to keep a company moving forward—tabulating sales and paying employees and shareholders. But change is a constant in this world. Those neglecting to heed the signs or respond to new realities often find themselves driven out of business. Every enterprise needs a road map for innovation—a plan to meet the opportunities of a changing world. This road map in a process-focused enterprise is called the *innovation plan.*

7

The Innovation Plan: A Methodical Approach Is Needed to Build a Skyscraper

The methodology and structures comprising a process-focused enterprise are unique in their ability to serve as a framework for innovation. Creating widespread awareness of the operational functioning of an enterprise and underpinning resource utilization onto this structure are momentous steps forward for most enterprises. The enhanced content and availability of critical information to formulate plans constitute a distinct break from the period when business plans were predicated on guesses as to how things worked and what could be changed. Armed with rich information and a thorough understanding of the enterprise's capabilities, the only remaining piece to build an innovation engine is a coordinated plan to take advantage of the available opportunities—a plan to marshal resources and make the right change happen. But today there are substantial impediments to executing this type of plan.

The prevailing corporate structure functions as a realm of fiefdoms where a local boss is the arbiter of where and when his or her workers will engage. Without their approval, these resources arc locked away until the chieftain is convinced that it is in their best interests to participate (and their voice is heard), or they are commanded to get on board by more senior leaders. More often than not, it is left to the initiative owner to convince the chieftains to participate by wooing them—an activity that is immensely time-consuming and wasteful, forestalls the initiative's delivery, diminishes any existing momentum, and may well erode any market advantage to be gained.

Although the process structure and governance organization diminish the silos in an enterprise and erode the power of the chieftains, improvement initiatives will not occur in a coordinated or timely fashion if leaders and workers are not executing from the same marching orders—in effect, a road map plotting the work to improve the position of the enterprise. This road map puts to work the theory of the four facets—capturing consumer insights, developing strategic and operational improvement initiatives from these insights, and then coordinating the efficient execution of these initiatives. The resulting plan I call an *innovation plan—a road map to transform an enterprise's processes to achieve strategic and operational goals while simultaneously maximizing the total value of its portfolio of improvements*. Extrapolating from this definition, effective innovation plans are born from a systematic approach that drives collaboration and stages adjustments to the overall process system in the most effective manner. And as mentioned previously, using process outcomes to communicate adjustments to the enterprise's operations fosters a universal understanding of the intended changes—uniting all enterprise resources behind a single improvement agenda. In short, it gets the congregation singing from the same hymnal.

WHAT IS AN INNOVATION PLAN?

In contrast to most companies—which communicate strategy based on leadership directives, documented metrics, or corporately approved goals—a process-based approach translates strategic intent into process adjustments. In other words, every improvement initiative can be linked to one or more processes, existing or new, and the initiative is defined by its impact on those underlying processes. To review the role processes play in innovation, *processes*, as defined previously, are simply activities that use inputs to produce outputs. The outputs possess attributes that either appeal to or repel a potential consumer. Every attribute of an output depends on the process's inputs and the process employed to transform them. Therefore, the easiest and most straightforward manner to produce desired attributes in an output is (1) to adjust the inputs or (2) to change the process employed to create the output. Both tactics require a deep and accurate understanding of the underlying process.

It is in this relationship between intent and process that the awesome power of process management lies. In popular business theory today, strategic execution and process improvement are often fallaciously viewed as disconnected concepts. In reality, process is the actualization of strategy. It is impossible to deliver a consistent customer experience (a strategy) without the execution of repeatable and specific processes. Viewing this relationship from a strategic standpoint allows for the discovery of an important truism: anticipated consumer needs and desires can be translated into specific output attributes and mapped to an enterprise's processes. The output attributes are the future requirements for processes—a direct connection between the customer's desires and the enterprise's intended future state. This link between the future and the current state of operations takes the guesswork out of strategy and innovation. You simply build what the customer is telling you to build. There is no easier way to strategically calibrate—and best yet, it does not require shelling out millions of dollars to get the latest and greatest advice from strategy firms.

The straightforward and relatively simple language of process adjustment belies the potential challenges of actually adjusting processes to customer wants and desires. First of all, processes rarely, if ever, are performed in isolation to deliver outputs. They are almost always a piece of a larger network of interconnected processes that collectively produce the outputs—in other words, a process system. Adjusting a process system to achieve a desired outcome requires an approach that weds collaboration across processes and their owners in a methodical and specific manner. When improvements are not designed and implemented at a system level, the solution may well fail to fully consider all the interconnected processes and thereby create a ripple effect of unforeseen consequences—possibly detrimental to other areas of the system. Although completed with the best intentions, the change resulting from such localized improvements may fail to deliver the intended outcome at the system level—the level visible to the consumer. The key to effective innovation is to create a comprehensive plan that considers opportunities to collaborate, cooperate, and share knowledge and resources with the goal of optimally allocating resources to objectives at the right time to maximize the value generated by the total portfolio of improvement initiatives. This is the aim of the *innovation plan*.

CONSTRUCTING AN INNOVATION PLAN

Although the leadership council is technically accountable, an individual or small group is commonly appointed to facilitate creation and management of the innovation plan. Regardless of the creator(s), the involvement of process sponsors and process owners is imperative to make the plan as complete and accurate as possible. The approach to building an innovation plan consists of five steps:

1. Select initiative-prioritization criteria.
2. Identify the process requirements for all known initiatives.
3. Build a strawman solution for each initiative, and compile the benefits, costs, and other information to allow for a fair comparison among initiatives.
4. Prioritize the initiatives to maximize the collective benefit for the enterprise.
5. Schedule and allocate resources according to the prioritization of the initiatives.

These five steps are intended to be executed sequentially, but in an ongoing iterative manner. For example, when scheduling and allocating resources, new resource requirements may be uncovered, forcing a project team to backtrack and recalculate the business case. Or, as limited resources are allocated, other initiatives may be delayed because of an overlapping need for the same resource. Still other initiatives may be less encumbered and therefore leapfrog these stalled initiatives. Additionally, an option when resources are scarce is to buy or borrow the needed resources, eliminating the restriction and the need for any adjustment to the prioritization order. When borrowing resources, an analysis of the benefit gained versus the cost of the borrowed resources is essential.

Step 1: Selecting Initiative-Prioritization Criteria

Prioritization criteria are any factors used to evaluate and rank initiatives, including the benefit delivered, the time to payback, the investment required, risk, and others. Selection of the criteria and the method to prioritize initiatives always should occur upfront and ideally even before the initiatives are known. Delaying this activity until the initiatives are bundled and ready for prioritization invites political haggling. Optimization of the portfolio heavily depends on the integrity of the initiative evaluations.

Today there is nothing close to a standard approach to ranking initiatives. Strategists argue that initiatives with a customer impact should take precedence because they are critical to future sales. To some extent, this is correct; the enterprise that loses focus on its customer faces a rocky future. Many leaders will argue that initiatives with a quick payback should take precedence because they build momentum. Other theorists argue a similar line and suggest focusing on initiatives that are easy to complete. My experience is that absolute rules such as these work some of the time, but just as often they fail to arrive at the correct answer.

Experience has taught me that there is no single approach that is applicable in all circumstances. The criteria to rank initiatives vary based on the financial, strategic, and operational state of the enterprise at a given point at time. For example, a cash-starved company with a large loan payment on the horizon may elevate initiatives requiring minimal investment but with very positive short-term outcomes (e.g., a cost-reduction initiative). Similarly, a company in an intensely competitive marketplace may focus on initiatives to build strategic advantages while forgoing short-term quick-fix initiatives.

In the absence of special circumstances, the main criterion for initiative prioritization should default to the expected net benefit of an initiative. This is best calculated as the *net present value* of an initiative. After all, the reason why enterprises exist is to generate value for their stakeholders. It follows that the most efficient portfolio is the one that maximizes the value delivered by the full portfolio of initiatives—both strategic and efficiency based. This approach ensures that valuable infrastructure investments are not ignored. Many companies, particularly retail and CPG (consumer packaged goods) companies, shoot for the shiny object versus the continued care and feeding of the existing infrastructure. Eventually, things begin to fall apart to the detriment of customer experience, which has a negative impact on customer loyalty and sales.

With value generation as the preeminent criterion for ranking initiatives, every initiative needs a business case. However, when value generation is the sole criterion, any possible launch order for initiatives needs to address the resource requirements and dependencies of the initiatives. If resources are not available or needed events have not transpired, an initiative cannot move forward. Simply waiting for these requirements to be available is wasteful. A better option is to shift resources and focus to initiatives that are ready to be executed and eliminate any time delay. To rank initiatives, the initial step of a prioritization process is to understand and capture for each initiative three vital pieces of information:

- *Value creation* (also called *net benefit* of the initiative) is the primary factor in nearly every prioritization exercise. As an enterprise generates greater value, it enjoys a corresponding financial gain, which allows it to meet financial obligations and invest in new capabilities. When value creation takes a back seat to other factors, an enterprise's future prospects are lessened. Only in extreme instances should other prioritization factors be ranked above value creation.
- *Resource requirements* are the people, money, and other resources necessary to execute an initiative. If the required resources are not available, the initiative cannot be completed. On occasion, shortages can be mitigated through the use of external resources, such as using consultants to fill knowledge gaps.
- *Dependencies* are prerequisites to the execution of an initiative. They come in a variety of forms including the occurrence of an event, an output from another initiative, a signed contract with a business partner, or anything needed to start an initiative.

After these three items are known, there are a nearly infinite number of potential prioritization criteria that the team can use to rank initiatives. In general, it is wise to limit the number of criteria. In most cases, the business case with consideration of resources requirements and dependencies is sufficient, but special circumstances may warrant the inclusion of other factors. The following criteria are occasionally considered for inclusion:

- *Significant known issues* are flaws that affect the consumer or hinder internal efficiency. If they create a negative experience for the consumer, the situation could develop into a competitive disadvantage and drive customers to explore substitute products/services. These issues usually can be accommodated in a detailed cost-benefit analysis. When the issue is not resolved, it creates an opportunity cost. Significant known issues are often addressed promptly because of their potentially large detrimental outcomes (e.g., the environmental and financial impact of an oil leak or the release of harmful gases into the air).

- *Strategic initiatives* that are vital to the organization's future market relevance can affect the cost-benefit in much the same manner as significant known issues. They generally drive revenue, and there is commonly an opportunity cost (especially with regard to lost market share) if they are not launched. This happens because a competitor may seize the advantage and make a move that thwarts the enterprise.

- *Customer/business-partner impact* initiatives are handled in the same manner as strategic initiatives. Similarly, they may have a benefit and an opportunity cost. Like significant known issues, they frequently have a meaningful opportunity cost if they are not addressed. On occasion, customer impacts may limit the work an enterprise can do.

For example, a partner might not have the capability
to engage in a joint improvement initiative because of a
noncompete agreement.

- *Risk mitigation* affects the business case of an initiative
and rarely needs to be considered separately. To account
for risks, a net benefit can be multiplied by a risk factor
to arrive at a risk-adjusted benefit. Risk might be reflected
as an opportunity cost. In other words, the enterprise may
suffer consequences if an initiative is not completed. A good
example of risk mitigation is compliance with regulatory
requirements. If the company complies, it avoids a fine. The
fine and the likelihood of the company being caught are
the risks the company willingly accepts and, in such a case,
represent the potential financial benefit of noncompliance.
The cost is the financial impact of complying. It sometimes
surprises leaders to discover the relatively low-risk weighted
impact of noncompliance.

- *Ease of execution* of the initiative is a factor that is
commonly used to prioritize initiatives. An argument is
often made that gaining momentum through a successful
launch of an initiative will position the enterprise for future
innovation efforts. When used as a prioritization factor,
ease of execution boils down to whether an initiative is
an incremental or transformational change (i.e., a cultural
change is required because of a radical change in mind-sets
and behaviors needed for it to succeed).

- *Size and scope* overlap slightly with ease of execution but
are focused more on the number of business units, vendors,
and partners; the time period; and resource requirements
of an initiative. The size and scope of an initiative can
be accounted for through resource requirements. These
requirements not only drive cost but also consume the lion's

share of an enterprise's available resources. An initiative with a large scope or size may delay the launch of other initiatives by draining the resource pool. Additionally, size and scope often go hand in hand with complexity. In general, any enterprise is limited in the number of complex initiatives it can effectively execute at any one time. Initiatives with overlapping scopes create even greater challenges.

- *Timing* as to the duration before the benefits are realized is addressed in the business case. It is standard to use a multiyear assessment and apply a discount factor to adjust the cash flows to present value in order to fairly compare different initiatives. On occasion, an enterprise requires a quick return on investment. This may be accomplished by prioritizing initiatives with short payback periods.

- *Cash impact* is unique in that it may become the most important prioritization consideration when an enterprise is faced with cash-management problems or when it is hoarding cash for a significant outlay.

To identify the right prioritization criteria, always start with the basics: net benefit generated, resource requirements, and initiative dependencies. The overwhelming majority of enterprises would stand to benefit by using only these three factors. But we know the world constantly changes, and the enterprise will go through periods when other criteria are important. On a somewhat regular basis, the prioritization factors and their reason for inclusion should be reevaluated to determine their relevancy. Make it simple. Using a straightforward and simple approach makes iterative reassessments that much easier.

After the initial three criteria, be selective in considering additional criteria. A long list introduces unnecessary complications, including the likelihood a criterion will be misapplied or incorrectly calculated, which leads to prioritization errors. In general, choose only factors that are applicable to the current environment and not reflected in the net benefit calculation. Simply adding factors because of convenience or their availability is a poor practice. For example, if a company is experiencing a significant volume of customer issues, it is logical that an initiative to resolve those issues would be a priority. There may even be a prioritization criterion called *customer impact* to reflect the leadership's concern with the magnitude of the issue. In reality, though, the inclusion of a customer impact factor is unnecessary. The business case should reflect the risk of a loss in sales and customer loyalty if the situation is not remedied. These costs (and other identifiable opportunity costs) should adequately incorporate the magnitude of the issue. If the customer impact is significant, the initiative will leap to the top of the execution order—that is, assuming that there are not other initiatives forecasting even greater value.

This example illustrates a situation that is common with many of the criteria listed earlier, including risk, strategic initiatives, and customer impact. If the correct benefits and opportunity costs are included in the business case, their impact goes directly to the bottom line. In most instances, further prioritization criteria outside of net benefit are unnecessary.

There are a handful of exceptions to this rule. As mentioned previously, cash requirements (i.e., investment required) occasionally take precedence. For similar reasons, the timing of an initiative's benefits may take precedence because of commitments made to debt holders or investors—especially Wall Street. One final exception is combating market intrusions. To preserve market share or a competitive position, companies may make investments with a minimal financial upside but with the intent to block a competitor from gaining a strategic advantage—in other words, the initiative would make the advantage less appealing to the competitor. An example is when a company buys the assets of an ailing competitor to prevent another enterprise from entering the market.

Once the prioritization criteria are selected, the leadership team needs to rank the criteria so that the initiatives can be analyzed and ordered for execution. As I have stated repeatedly, the primary consideration in almost all prioritizations should be the value generated by each initiative. However, in instances where additional criteria are included, another step is necessary: the prioritization factors need to be ranked in order of their importance to the enterprise at that moment in time. This becomes a governing assumption of the prioritization process; if the ranking of the criteria changes, the portfolio's initiative order will need to be reassessed. In most cases, when there is an additional prioritization factor, it will take precedence in the short term or until the conditions for which it was introduced are mitigated. For example, a company may have a large cash outlay as a result of a lawsuit. Cash requirements then would become the primary prioritization consideration. Initiatives may be launched as long as they have minimal cash requirements. Once the payment is made, the company's cash position improves, and cash requirements can be eliminated as a prioritization factor. This, in turn, launches a reevaluation of the portfolio. Net benefit is again the primary factor, and the initiative launch order is adjusted to reflect the new emphasis.

After selecting the prioritization criteria, leadership teams may wish to consider more than one factor. In this case, an algorithm can be created to weight every prioritization factor according to its importance at that point in time. As circumstances change, the relative weighting of the factors is adjusted to reflect business priorities. With the prioritization criteria selected and ranked, the next step in the prioritization process is to identify the slate of initiatives and their process requirements.

Step 2: Identification of Process Requirements for All Known Initiatives

Compiling a list of the intended process requirements for initiatives is not always a straightforward or simple exercise. Only in rare instances does a company maintain a full list of active and queued-up initiatives, and almost never is an end state available for every initiative. Even though it is very common for companies to operate some form of program management office (PMO) to monitor progress on major initiatives, these groups generally focus on strategic initiatives to the detriment of operational-improvement initiatives. And only frequently are initiatives logged with a consistent format and sufficient definition to prioritize their execution. Thus a good amount of digging and translating is required to pull a list together.

The first step is to identify the major initiatives. As defined previously, an *initiative* is a project or a group of projects that, when viewed together, achieve a goal. The list of proposed initiatives generated from the strategic planning process and operational improvements is an unorganized but quick view of an enterprise's business plan. It is broader than a strategic plan because it incorporates operational-improvement initiatives—hence it is correct to label it as a *business plan*. Although a high-level view of each initiative exists, at this point the initiatives remain rather vague and subject to interpretation. Like a strategy, initiatives are most effective when they are specific and of manageable size but still deliver a benefit. Logical delineations for dividing work efforts into individual initiatives are factors such as geography, different product/service lines, unique customers, and timing.

When forming the list, the intent is not solely to identify the specific initiatives but to gather all known information. Not uncommonly, there are gaps because poorly defined initiatives are the norm. Many initiatives (especially newer ones) exist as empty vessels with insufficient detail to even discern their true intent. On the flip side, reams of information are frequently available for major initiatives —truckloads more data than is necessary, but is it the right information? To put a stake in the ground, thorough communication of an initiative requires an initiative name, owner, objective, scope, timeline, benefits/business case, assumptions, risks, resource requirements, high-level work plans, and any related costs incurred previously or anticipated in the short-term future. If the information is not immediately available, that is fine. In all but the most rare instances, initiatives require further analysis and design to prepare them for inclusion in an innovation plan. Initiatives fall into one of two buckets—strategic initiatives or operational-improvement initiatives. Both are integral to an innovative enterprise. The difference lies in the manner in which they are identified and managed.

When capturing a list of initiatives for the first time, the optimal time to start is immediately after the conclusion of a strategic planning cycle. The strategic plan includes many of an enterprise's major initiatives, making it a convenient source when compiling a list of initiatives. But strategic plans very seldom include the full set of strategic initiatives. Companies continuously launch initiatives to counter competitive threats, to course correct in the wake of dismal financial results, or to respond to a new customer trend. Although seemingly logical to do so, few enterprises manage an ongoing improvement portfolio in the periods between strategic planning cycles. When this is the case, the quickest route to uncovering midyear strategic initiatives launched outside a strategic planning cycle is to survey functional leaders to whom these initiatives were pushed for execution.

After compiling the list of strategic initiatives, the next activity is to capture the operational-improvement initiatives. Operational-improvement initiatives result mainly from growth and adjustments to the enterprise's market focus. Many of the initiatives in this bucket trace their routes to a prior year's strategic plan. Either while designing the initiative or after its launch, it is discovered that the strategy requires capabilities not currently in existence. As a result, investments are made to expand the infrastructure and develop the needed capabilities. Such investments appear in the form of technology upgrades, new facilities or machinery, personnel growth, or the introduction of procedures/functions to comply with government regulations. To identify operational-improvement initiatives, examine the current and prior strategic plans. Look for investments made as a result of strategic endeavors. Often these investments are made to supporting functions to increase their scalability. An equally good method is to follow the money. Look for major expenditures in the budget and general ledger, especially capital expenditures.

Other operational-improvement initiatives are often void of any monetary investment. Indeed, they may require little to no investment because the individuals staffed to complete the initiative are lent from other business teams. And unlike strategic initiatives, it is rare to find a company that tracks operational-improvement initiatives at an enterprise level. However, there are telltale indicators to dig them out. Operational-improvement initiatives often require specialized expertise (e.g., Six Sigma or Lean) to run them. For this reason, tracking the spending on consulting services and training is a good way to uncover operational-improvement initiatives. A related strategy is to locate the individuals with expertise in improvement methodologies such as Lean or Six Sigma and identify where they are spending their time. Endeavors using improvement methodologies or skilled resources are either strategic or operational-improvement initiatives. The type of initiative is inconsequential—identification of initiatives is our goal.

As an aside, portfolio-management organizations often disregard the tracking and management of operational-improvement initiatives because of the lack of a significant investment. This is a mistake. Resources, especially skilled and knowledgeable employees, are limited in most environments. An efficient portfolio accounts for and manages all types of limited resources—not just the dollars.

At this point, a laundry list of initiatives is assembled—albeit a list lacking any evaluation of the legitimacy of any single initiative. The list is often overly large and contains initiatives of vastly different scopes and anticipated benefits. In many cases, the list includes initiatives that are contradictory in their stated objectives. And the list will have some real dogs. Dogs are initiatives slated to deliver either no benefit or negatively affect the enterprise. They continue to limp along, in many cases, because their utility is never reevaluated once the initiative is in progress.

The intent of an innovation plan is to take this full lineup of initiatives, analyze them in a methodical manner, and then prioritize their execution to maximize the total benefit generated. To kick-start the process, conduct an initial culling to verify that every initiative meets a basic set of criteria—e.g., scope, value generated, or resource usage—to be included on the list. However, conduct even this first round of eliminations with care. The available information is based on prior evaluation methods. Different people used different approaches to design the initiatives—potentially resulting in biases or inaccuracies that might lead to the elimination of a valuable initiative. Use care, but recognize that at least some of the initiatives are unworthy of further exploration.

Innovation planning is the approach used to evaluate, execute, and deliver improvements with a cross-functional scope that require a significant investment of resources or are critical to future success. The criteria for the cutoff line vary based on the goals of the enterprise and current financial circumstances:

- Value delivered by the initiative is usually a primary consideration. Including an initiative that fails to generate value runs counter to the goal of an innovation plan.
- Size, measured by investment, scope, and cross-functional reach, is another potential differentiator. Large-scale initiatives with significant investments need to remain in the portfolio. On the flip side, an initiative may be small and seemingly insignificant, but it is integral to the achievement of larger strategic goals. They need to be in the portfolio.

However smaller, independently executed initiatives with minimal investment of resources are candidates for exclusion because they can be completed without the coordination of cross-enterprise resources. In other words, they are delegated to a process owner for execution—not abandoned.

■ Impact is another differentiator. Even if the initiative is small in scale, it may provide great benefit to the enterprise and be worthwhile to track at an enterprise level.

■ Dependencies with other initiatives are a prime consideration because work efforts need to be coordinated.

In general, keep the larger, complex, interconnected, cross-functional initiatives in the innovation-planning process and push the remainder to the appropriate process owners for execution or elimination. To prevent the formation of a shadow improvement portfolio, clear guidelines should be established to segment initiatives to the appropriate place. Limited-scope improvements managed at the process-owner level are intended to improve the base-level process but not impact or diminish larger enterprise innovation efforts or even affect areas outside the immediate process. For those improvement efforts outside the innovation plan, the intent is to provide some measure of attention for every significant process and further, that when process teams have excess capacity, they focus on smaller improvement efforts.

A final consideration of this initial evaluation of the initiatives is to weed out initiatives that overlap or do not make sense when both are implemented. This often occurs with strategic and operational-improvement initiatives aimed at the same area of the enterprise. The strategic initiative fundamentally alters the output of the process. The efficiency initiative delivers the old output at a lesser cost. As a general rule, always complete the strategic work before any efficiency work. This prevents the wasteful improvement of a process that is slated for retirement. Thus, whereas there are no hard rules, an initial culling of the portfolio weeds out initiatives that are insignificant, duplicative, small in investment and scope, or without benefit.

Beyond an Initiative: To a Solution

One of the greatest challenges in launching initiatives is to overcome the lack of definition or clarity around their intent. It is rare to find sufficiently detailed initiatives that can move forward without a good amount of interpretation. With vagueness being the standard, attack the ambiguity by sketching out the initiative's end result. This sketch does not have to be overly detailed. In fact, little more than a strawman is needed to build a sufficiently clear picture. Of course, after resources are allocated and the initiative is launched, the design may be adjusted or expanded by the initiative team.

An individual or team usually leads the strawman design. For smaller initiatives, the owner (functional or process) of the area most affected by the initiative is best positioned to lead this work. For cross-functional initiatives of moderate size, the leaders may be one or more managers working in tandem—often the process owners of the areas most affected. For initiatives spanning multiple areas and with major complexity, leveraging the skills and knowledge of a team of subject-matter experts is the best approach. In this type of situation, process facilitators often lead the team because they bring objectivity and an enterprise-wide perspective.

Capturing the intent of an initiative requires a return to its roots. Begin by interviewing the initiative's originators to get a clear, firsthand perspective. However, the initiative may not have sufficient clarity even after these interviews. At the end of day, use the best-available information to sketch out a rough outline of the intended result. The exercise is not to build with precision in mind but rather to provide sufficient detail to estimate the costs and benefits of the initiative.

Make it easy. A solid solution design is a paragraph that identifies the work to be completed and the proposed end result. The solution design communicates the scope of the effort, lists customer connection points and the anticipated end state, identifies key stakeholders and performers, and details the differences between the current state and the future state. *Proposed* is a key word for the design. Once the initiative is started, it may need modifications as new information surfaces. This is why initiative management is always an ongoing exercise. The world changes, and to achieve success, so must the initiatives.

For example, the following solution design was written for an initiative aimed at improving the distribution processes at a regional grocery chain:

> The intent is to optimize the number of distribution centers (DCs) from a cost perspective while accommodating a two-day delivery to customers' receiving facilities. In the initial stage of the initiative, the focus is on analysis of the current structure and determination of the optimal transportation routes to meet the delivery goal. With the knowledge gained in this initial step, the team will plan the DC network and determine where new DCs should be located, which DCs should be maintained or adjusted, and which DCs will be shuttered. The following phase of the initiative will include the mobilization of a team to implement the new DC arrangement. Scoping and building this initiative require a collaborative effort including resources from store operations, real estate, and the supply-chain team. The internal operations of the DCs are not anticipated to change other than to accommodate the new distribution routes.

Once a solution design is complete, always circulate it to the affected process owners (or functional leaders) as well as the initiative's originators for their review and confirmation of the solution. This validation step confirms that the solution is viable and consistent with intentions.

Process Requirements

Initiatives are executed under the premise that on completion, the enterprise will realize a benefit. In corporate America, many initiatives fail to meet this simple benchmark for a gazillion reasons, including a lack of commitment (exemplified by inadequate resource allocations), poor solution designs, incomplete execution of the initiative, and others. That said, a good number of initiatives are destined to fail before they leave the gate because their intent is fatally flawed. The initiative fails to target the right customer or prepare for competitor reactions. In short, the initiative is strategically defective.

Even if an initiative is incontrovertibly dialed into the winning strategic position, a major speed bump remains—the precise communication as to what is to be accomplished and how it is to be done. Aside from poor design, the most prevalent failure point is the widespread inability of strategists to transcribe their ideas in a format that is understandable and executable. Vagueness leads to misinterpretations, guesswork, and waste. To have a prayer at linking strategy to execution, initiatives need to provide not only the end state but also direction on how the initiative is to be completed. Leaders and managers everywhere struggle with this challenge—giving rise to continual theories and tools to bridge the gap. Until now, though, the answer has been elusive. Fortunately, there is a simple and straightforward way to define initiatives—a way to forge an iron link between execution and strategic intent. This is accomplished by translating initiatives into *process requirements*.

What are process requirements? Process requirements identify exactly what is needed from a specific process to deliver the desired outcome. It breaks down the desired outcomes into components contributed by different processes. Process requirements communicate with exactness not only the intent but also the "How?" of an initiative.

For example, consider the proverbial widget. Based on customer and market analysis, a company identifies an opportunity to sell a 4-foot widget with a floater value. This new product is predicted to be a game changer, and therefore, speed to market is paramount to success. We start with the end result—a new 4-foot widget with a floater valve ready for market. What is it going to take to launch this product? Because it is not an existing product, research and development (R&D), engineering, manufacturing, and procurement will need to design the product, build the manufacturing capabilities, and procure the raw materials for its production. Then marketing and sales will need to build awareness and line up customers. Finally, the supply chain will need to transport the new product to the customers' locations. The process requirements for this initiative might be documented as follows:

Four-Foot Widget with Floater Value

- *Customer analytics.* Gather information on the desired customer attributes; in other words, build customer specifications for the 4-foot widget.
- *Strategy.* Assess competitor reactions to the launch, and create a market strategy for the new product.
- *R&D.* Design and develop a 4-foot widget with a floater value based on customer specifications.
- *Engineering.* Create the machinery and other production processes to build the 4-foot widget.
- *Procurement.* Acquire the raw materials to manufacture the new product.
- *Manufacturing.* Train and prepare workers to produce the new product.
- *Marketing.* Develop the packaging, pricing, and promotional program to launch the 4-foot widget.
- *Supply chain/distribution.* Assess and prepare for distributing the new product to distributors and customers across the target market.
- *Sales.* Identify the customers and build a sales organization to support the product launch.

Other process requirements arguably could be added to the list. One omission in this example is the lack of supporting-process impacts—that is, information technology, human resources, and finance. Whenever possible, existing processes should be leveraged for new-product innovations in order to reduce the delivery time and overall costs. Of course, over time, growth may stress the enterprise's infrastructure, necessitating eventual upgrades to continue producing and selling a product.

Returning to the list of initiatives, capturing the process requirements is as simple as going initiative by initiative and identifying the processes impacts. To ensure completeness and accuracy, process requirements are best determined by the individuals with the greatest familiarity with existing processes and their adaptability to a new use. This is why it is imperative that process sponsors and process owners are engaged in this exercise. To facilitate the collection of process impacts, it is helpful to use a table similar to Table 7.1

In this table, the company is focusing its attention on two major initiatives. First, it plans to launch a nutritional rating scale for its products to aid health conscious customers. Second, the grocery chain intends to open six new stores. The process requirements for the megaprocesses were derived during a design meeting with functional leaders. Every process requirement can be thought of as a small project supporting accomplishment of the initiative. This table can be further expanded beyond megaprocesses to show the impacts on additional levels of the process structure (i.e., major processes, processes, and subprocesses). Usually the major processes are at a sufficient level for scoping, and further decomposition can be delayed until after the initiative team is onboard.

Building a table similar to Table 7.1 is a relatively straightforward process. To identify the impacts, the process sponsors (or suitable representatives) convene to review the portfolio of initiatives. The team discusses the details of each initiative, and as each initiative

Table 7.1   Process Requirements for Initiatives at a Regional Grocery Chain

	Megaprocesses/Functional Areas			
Initiative	Marketing	Merchandising Operations	Supply Chain	Retail Stores
Launch nutritional scale for products	Communicate to consumers Create signage/product information	Develop process to assess product nutrition (accurate and complete)		Place labels and materials to support nutritional scale Train workers on program
Build six new stores	Create awareness of company in new markets Host grand opening	Develop/manage vendor relationships to support new stores (minor)	Develop routes to support delivery to and from new stores Supply inventory to the new stores	Support opening of new stores by expanding infrastructure Hire and train new workers Set new store

is reviewed, the attendees explain the impact it will have on their areas—stating in process terms what needs to be adjusted, executed, or built. If an initiative does not affect a particular megaprocess, it is left blank in the table. Processes that are minimally affected by an initiative may list the impact but denote it as minor. The team reviews all the initiatives in this manner. A significant benefit of this exercise is that not only does it capture process requirements, but it also builds universal awareness of upcoming initiatives and facilitates the identification of collaboration opportunities.

During the discussion of each initiative, the team occasionally discovers that the enterprise is missing critical elements or capabilities to complete the initiative. A process may not exist that is capable of producing the desired output. A new process or perhaps even a new business model is needed. In most cases, the solution is simple: build a new process within the confines of the current process structure. Returning to the example in Table 7.1, merchandising operations is tasked with creating a new process to assess the nutritional value of different products. Although this is a new process, it logically fits under the umbrella of the merchandise operations organization. This is the simplest route to providing a missing capability. At the other end of the spectrum, when an initiative requires a unique business model, the team may have to build a new enterprise process blueprint (i.e., expanding into a new product line or into a new geography) and its supporting processes. Whenever possible, it is always more efficient to leverage existing processes.

On occasion, all of an initiative's process requirements may not be known or even available during the initial pass. This is especially true for new product lines or when the enterprise expands into new businesses. For example, consider the strategic goal of crossing a geographic border and expanding into a new country. The initial phase of such an initiative might be to assess the business practices unique to the new country. Entering a foreign market entails compliance with additional regulatory institutions and potentially different accounting rules—not to mention respecting local customs and abiding by local business practices. As is often the case when entering a foreign market, local experts might be hired to shepherd the market entry. Until an initial assessment is complete, the process requirements are unknown. In instances such as this, take the scoping and estimation exercise as far as possible. Include in the analysis what details are known, and create estimates for the unknown. As long as a consistent approach is used to estimate and rank the different initiatives, the innovation plan will be as efficient as the information allows.

Unlike other methods used to define initiatives, process requirements explicitly identify the processes affected and the expected outcomes. This approach to defining initiatives jump-starts their launch in a number of ways. Primarily, it provides immediate perspective as to an initiative's intent and scope—removing any guesswork. Second, it identifies the affected processes, their roles, and the expected outcomes for the initiative—setting the stage for collaboration between business partners. Finally, it sets a foundation for understanding the resource requirements and costs of moving forward with the initiative—acting as an input to a resource-allocation process. In an innovative enterprise, process requirements are a necessity for crisp communication.

As initiatives progress, the business case and other details begin coming to light. Before getting too deep into the execution of initiatives, it helps to build rules around their ongoing management—especially in regard to their prioritization for execution. The next step in building an efficient innovation plan is to use identified prioritization criteria and collect this information for each initiative.

Step 3: Collecting the Prioritization Criteria for Each Initiative

The tabulation of prioritization criteria builds on work already completed. The business case for each initiative is undoubtedly the most important piece of the prioritization process. Initiatives are undertaken because it is believed that they will generate value (except in rare instances). This is where the initiative launch process starts for many companies today.

Business Case (Cost-Benefit Analysis)

- The business case is defined in terms of an initiative's net benefit, which is the sum of benefits and costs.
- Benefits are the anticipated financial gains after deployment of an initiative, and they are measured over a set period of time. Three- or five-year time horizons are recommended because the accuracy of predicting benefit streams deteriorates as they extend into the future.
- Costs are expenditures required to capture the benefit—both during the course of completing the initiative and on an ongoing basis as the benefits accrue.

If an initiative forecasts a respectable net benefit, more often than not resources are assigned and the project is launched. However somewhere over the years, the real advantage in calculating an initiative's business case got lost in business practice. The most important usage of the business case is to appropriately prioritize initiatives for action. The comparative value created by each initiative is the critical piece of information needed to build the optimal value-generating improvement portfolio. To make this analysis work, consistency is key. Accurate prioritization requires an apples-to-apples comparison. Any comparisons fall apart if different techniques are utilized to estimate the value of initiatives or if different numbers are used for financial constants such as discount factors or if the team incorporates varied levels of detail into the initiative estimates. Achieving a fair comparison requires consistency across all aspects of the analysis. This means using the same calculations, taking the analysis to the same level of detail, and using the same constants (i.e., discount rates, corporate benefit percentages, etc.).

When building business cases, consistency is slightly more important than overall accuracy. In the event that a single financial assumption is off, the number skews all the initiatives where it appears. Depending on the relative sensitivity of an initiative to a constant, the rankings may be slightly affected. However, as more accurate data become available, the financial calculations can be adjusted during ongoing assessments, and when circumstances dictate, updates can be made to an initiative's ranking. The intent is to make the best guess as to which initiative provides the greatest benefit at a particular point in time. And it is a *guess*—not an absolute number by any stretch.

Because consistency is the key, ownership and responsibility for the process to generate business cases should be assigned to a single financial planning group. This group owns the methodology for building business cases and ensures consistency in the handling of each initiative. Additionally, this group is responsible for creating estimation mechanisms and tools to make the process more accurate and efficient. In this way, the group's role is like that of any other process owner—manage and improve the processes in their domain. Over time, the repeated use of a common approach creates an institutional proficiency—the accuracy of estimates increases, and the turnaround time to produce the estimates should decline. Process-focused enterprises enjoy an advantage out of the gate in estimating business cases because the use of process requirements allows for easy identification of all the affected parts of the organization.

Painting a complete financial picture of an initiative entails the calculation of two pieces: the costs to develop the solution and the ongoing benefits and costs after the solution is fully implemented. The benefits normally come to fruition only after the project team is done with its work, and the end state is fully deployed.

Starting with delivery of the solution, the primary costs are the project team (human resources costs); investments in facilities, machinery, tools (information technology and others), and other assets; any contracts or agreements with service providers; raw materials; and basic project team expenses, including computer usage, telephones, printers, and so on. These costs are not always easy to identify. The key is to put estimates down on paper via brainstorming or by using historical cases as a guide. After a cost is identified, conduct a quick estimate of its size and applicability. A team could spend weeks identifying and incorporating every possible cost into the calculation, but this is unnecessary. In general, the determination of whether a cost is material resides in two questions: (1) Is the cost significant enough to affect the prioritization order of the initiatives? And (2) will the cost affect other initiatives and therefore need to be included for consistency? When determining whether to include a cost, the financial planning team owns the final decision. With insight into all business cases, the financial planning team is in the perfect position to arbitrate whether any cost is appropriate for inclusion and to act as a source for historical information on initiative teams and their size, composition, and other factors.

Often the hardest cost to calculate is the cost of human resources because it is not immediately evident like investments or other expenses. One useful method to estimate the cost of completing an initiative is to make an educated guess as to the number, time commitment, and level of individual(s) needed to bring the initiative's solution to fruition. Chapter 6 provides guidelines as to the composition of a task force. Depending on the unique circumstances of the initiative, additional roles, including project managers, change managers, trainers, and others, are likely necessary.

The next step is to sketch out a rough timeline. By knowing what needs to be done and resource requirements, an educated guess can be made as to the duration of the initiative from start to finish. With their knowledge of the subject matter, process owners are ideally positioned to make these predictions. Always take advantage of institutional experience and knowledge from prior forecasting exercises to improve the estimates. As a shortcut, use "High," "Medium," and "Low" labels in place of more exact estimates. For example, short-duration projects may last 6 or 12 weeks. Medium-duration projects may be chosen to be 24 or 36 weeks. Long projects may be 52 or 104 weeks. Of course, when actual timelines exist, they can be used in place of increments.

Knowing the resource requirements and the timeline leads us to a rather simple calculation for the human resources costs. Take the estimated project duration and multiple it by a standard rate for each team member (Table 7.2) to arrive at a total cost for each resource. In this example, an initiative requires two resources—one part time and one full time for six weeks. The full-time resource, or full-time equivalent (FTE), plays the role of a task force member. Assuming a 40-hour work week, the cost for the full-time resource is 6 weeks × 40 hours × $50 an hour from Table 7.2. The part-time resource is a Lean expert. The cost for the part-time resource is 6 weeks × 20 hours × $150. This equates to a total of $12,000 for the full-time resource and $18,000 for the part-time resource.

When estimating resource costs, resources with similar capabilities should use the same rate to allow for an accurate comparison. Avoid using actual rates. Because actual rates may differ between resources (i.e. some of them are expert negotiators) with similar knowledge and skill sets, the results may introduce a bias into the analysis. The exception to this rule is when a specific individual is required and no substitute is available. In this situation, use the actual rate.

TABLE 7.2   Schedule of Standard Resource Rates

Skill Set and Knowledge of Generic Resource	Rate
Expert (FTE)	$150 an hour
Process owner (FTE)	$75 an hour
Initiative team member (FTE)	$50 an hour
Other resources (FTE admin.)	$35 an hour

With resource costs identified, the remainder of the tangible costs for the initiative can be identified by process owners and other engaged parties. These costs include equipment, facilities, technology, and other hard costs. As the costs are tabulated, be sure to capture the method used to build the calculations as well as any specific constants used. This background information is invaluable when auditing the process for improvements, and it accelerates updating the business case as new information surfaces. In a similar manner, capture all assumptions behind the business case. In some instances, the financial projections are simply educated guesses based on available data. To account for variations in any cost item, use several estimates (e.g., good, better, best) when building the business case to calculate a range for the item. As data updates become available, the estimates can be refined, heightening their accuracy.

After addressing the initiative costs, the focus shifts to the costs and benefits of the solution post implementation. The methodology employed is the same as that used to calculate the costs of deriving the solution. In lieu of a project team, the solution is transitioned to individuals who manage and perform the process on an ongoing basis. Postimplementation costs and benefits are frequently volume based. In other words, as a cost driver such as sales increases, the costs to produce those sales will increase as well. Forecasting sales levels is always a risky undertaking. As a rule of thumb, be conservative yet realistic in the estimates of sales and other variables. Document the assumptions and the calculations, and be prepared to update them as new information becomes available.

Capturing all the costs and benefits of a solution is tricky. The goal is to include only the meaningful costs and benefits and to maintain consistency in the evaluation process. Brainstorm the outcomes of the initiative. List the benefits that might result, and then do the same for costs. Another tactic is to walk through the affected processes and consider the potential benefits. Is there a cost savings? Is additional revenue potentially generated? Poll the managers affected by the initiative. At each analysis point, collect the data to refine the cost and benefit calculations, document the assumptions, and validate the results with process owners and other stakeholders. Iteration begets accuracy. Rarely is all the information immediately available. When questions arise as to whether to include a benefit, the financial planning team is the arbiter of whether the benefit merits inclusion.

When building a business case, the question often arises as to whether to incorporate not only the hard costs and directly connected benefits but also the costs and benefits that are less tangible. This is a challenge for organizations that are driven by Wall Street's expectations and that routinely disregard soft benefits, including risk mitigation, knowledge gained, increased strategic flexibility, and others. However, if risk and opportunity costs are not included, the potential exists to overlook needed infrastructure investments—putting major revenue streams at risk when the infrastructure is fragile. As a general rule, when evaluating any single initiative, I recommend including the opportunity costs and benefits that are directly attributable to the initiative. In order to account for the likelihood that the conditions warrant the inclusion, assign them a risk weighting. This risk weighting is a percentage that reflects the likelihood that the benefit will be captured or that the cost will be incurred. For example, if there is a risk that the benefit will not be captured because of the potential entry of a competitor, the risk factor's magnitude will increase with the probability that the competitor will enter the market.

For example, let's say that the government implements a new regulation on working conditions for workers. Compliance with the regulation will cost the company $100,000. The fine for noncompliance is $5,000. On reviewing the regulation's enforcement mechanisms, a team estimates the risk of getting caught at 10 percent.

The expected loss is then $5,000 × 10 percent = $500. Because the cost for compliance is $100,000, it is a poor financial decision to comply with the regulation. But these are only the hard costs. In the event the company does not comply, workers will likely take their case to the press. The loss of goodwill could greatly hurt the company's sales because the market has ample substitute products. A team estimates the loss may be as much as 20 percent of the company's sales ($50 million) with a risk factor of 50 percent. This equates to an expected loss of $5 million in sales. Assuming a 10 percent net margin, the company expects to loss $500,000 in profit. In this example, when we account for possible customer attrition, the numbers point to a very different conclusion. Accounting for goodwill and any corresponding bump in sales is a good example of factoring soft benefits into a business case. With the passage of time, soft costs and benefits convert into real benefits and costs.

With all known costs and benefits baked into the business case, the net-benefit calculation comes down to math. Discounting the cost and benefit streams by the appropriate discount factor yields a net benefit for the initiative. Again, document all the assumptions and calculations made during the course of building the business case.

While compiling the business case, it is often convenient to capture other prioritization factors for the initiative. For example, if cash requirement is a prioritization factor, the team can estimate the project's cash needs when pulling together the business case. Again, all assumptions should be documented. With a solid draft of the business case, the next step is to focus on the interdependencies between the initiatives and their resource requirements.

Capturing Initiative Dependencies and Resource Requirements

To identify the other restrictions on an initiative's launch, begin by evaluating each initiative and listing its dependencies and resource requirements. Many enterprises forget this crucial step and eventually pay the price for their lack of foresight. It is common to find new strategic initiatives launched without considering precedents. By the time they are addressed and the initiative is ready for kick-off, the anticipated competitive advantage has been eroded or is completely gone.

Initiative dependencies and resource requirements are similar to each other. Both are precedents for an initiative, and both can derail an initiative's execution and render it worthless.

Dependencies are any activity or event on which the initiative depends and must be completed before the initiative can be executed. For example, a business partner might need to inaugurate a new leadership team before a contractual agreement can be formalized. Or the completion of the dependency might provide a resource needed for the initiative. For instance, the submission of incorporation documentation to a state agency must occur before a company can receive a tax ID number and set up a business banking account. Dependencies may be on other initiatives or on the completion of an external event.

Resource requirements are the specific inputs required to execute an initiative. The focus is primarily on resources that are needed but not readily available. They require time and frequently money to obtain. Limited resources include highly skilled individuals, machinery, facilities, the participation of specific process owners or business partners, and, of course, money. Operating in a process-focused environment greatly aids the collection of resource requirements because simply knowing the process requirements jump-starts their discovery.

In most instances, the simplest way to identify dependencies and resource requirements is to review each initiative and determine

- The logical starting point of the initiative.
- Any assumptions required for the initiative to begin at that point (i.e., the resources or inputs that must be available at the inception).

■ Other inputs needed throughout the duration of the initiative's execution (i.e., inputs required once the initiative is in flight). These resources may be personnel, equipment, information, an agreement to be confirmed, or even a decision to be made.

Inputs are the meaningful element. When required inputs are unavailable, work cannot proceed. If the input is an output of another initiative, a dependency exists on that other initiative. If a resource is needed, there is a resource requirement to be fulfilled. The prioritization of initiatives must reflect the initiative's needs for it to be a realistic and actionable plan.

Step 4: Prioritizing the Initiatives

The end goal of the prioritization exercise is an unbiased and accurate comparison of initiatives. Before beginning, give the leadership team time to review the innovation portfolio one more time. This reexamination serves as a final checkpoint to eliminate initiatives that barely made it through prior reviews and no longer make sense to move forward. Additionally, this final review incorporates the most up-to-date knowledge of the current environment and the enterprise's strategic focus—guaranteeing that initiatives unaligned with the strategic focus of the enterprise are jettisoned.

After paring down the list of initiatives, the prioritization process begins. For each initiative, we know, at a minimum, its resource requirements, its dependencies (as well as its synergies with other initiatives), and its business case. If other prioritization criteria are to be used, these details should be available for each initiative as well. To start off, it helps to consolidate the information into a single, organized list. Table 7.3 is a great example of a form that fits this purpose. This table allows for a quick view of the initiatives and highlights the details, including net benefit, dependencies, and resource requirements. Additional prioritization criteria, as well as risk factors, projected durations for each initiative, process impacts, and the current status of previously launched initiatives, are also included. When space allows, the assumptions made while developing the initiative can be listed.

A consolidated view provides convenient access to the pertinent information needed to rank initiatives. When it is in a database format, the first pass at prioritization is easy—just sorting the initiatives by the primary prioritization criterion and then completing successive sorts on any additional prioritization criteria.

On occasion, enterprises opt for a more numerical approach. Algorithms frequently include factors such as duration of the initiatives, risk factors, and other factors such as cost. If these factors are entered on a spreadsheet, an algorithm is easily created to derive the prioritization. With the initial prioritization complete, the result is the *ideal state* and represents the greatest potential value generation of the innovation portfolio. However, only in rare instances can the initiatives be executed in this order. As stated previously, many initiatives have predecessors that must be completed prior to their execution. And then the law of scarcity comes into play. People, knowledge, raw materials, expertise, and machinery—the unavailability of any of these factors potentially limits an enterprise's ability to start an initiative. Building a realistic execution order entails accounting for dependencies and the availability of limited resources.

Step 5: Scheduling Initiatives and Allocating Resources

Building the actual execution order is a two-step process. With the prioritization criteria incorporated into the initiative order, dependencies become the initial focus.

TABLE 7.3   Innovation Table Prioritization

						Resource Requirements	
Initiative	Risk Adjusted Contribution (000's)	NPV(000's)	Risk Factor	Design or Execution	Initiative Budget $ in 000's	Team	Special Skills
Customer Behaviorial Research Study	$1,400	$2,000	30%	Execution	$10	Process Owner	NA
Customer Relationship Management System Launch	$119	$125	5%	Design	$75	Task Force	CRM Specialist
New Offering Launch	$360	$400	10%	Execution	$60	Process Owner	NA
Working Capital Assessment	$45	$50	10%	Design	$15	Process Owner	Financial Analysis

Other equirements	Dependencies	Collaboration	Duration (mos.)	Major Process(es) Impacted	Delivery Status	Projected End Date
▪	None	None	1	Customer Acquisition New Product Development Marketing Customer Service	Green	10/25/2009
an	None	None	6	Customer Acquisition Customer Service	Green	3/25/2009
▪	None	None	6	Customer Acquisition Operations Marketing	TBD	TBD
▪	CRM System Launch	Financial Assesment	2	Accounts Payable Accounts Receivable Customer Acquisition	TBD	TBD

A pseudo–Gantt chart (initiative chart) like the one shown in Figure 7.1 is a convenient view for evaluating dependencies. The chart identifies the initiatives and their respective durations.

- Start with the initial initiative in the sorted list. Check whether it has any dependencies.
- If it is without dependencies, it retains the initial position in the order. Set the start date for the initiative in the current month (January in Figure 7.1).
- If there is a dependency for the initiative, examine the linkage between the initiative and the dependency.
  - ▲ If the dependency is an event that must precede execution of the initiative, place the initiative in the month the event is anticipated to conclude. The customer behavioral research study initiative in this figure is an initiative with a timing dependency completed by May.
  - ▲ If the dependency is another initiative later in the prioritization order, the next step is to determine where an initiative consisting of the current initiative and the dependency would fall in the prioritization order. To do this, the prioritization criteria need to be calculated or developed for a joint initiative (i.e., the initial initiative and the dependent initiative). If value creation is the

FIGURE 7.1 Initiative deployment chart.

Initiatives	Jan	Feb	Mar	Apr	May	Jun	Jul	Aug	Sep	Oct	Nov	Dec
Customer Behavioral Research Study												
Customer Relationship Management System Launch												
New Offering Launch												
Working Capital Assessment												
TBD...												

primary prioritization criteria, the financial costs and benefits of the individual initiatives should be summed, and then the net present value of the joint initiative can be calculated.

■ Once this is done, the new initiative can be prioritized appropriately. This is accomplished by resorting the remaining initiatives to determine where this joint initiative falls. When this sort is complete, continue the process with the new top initiative.

Repeat these steps until all the initiatives are ordered based on the prioritization criteria and accounting for dependencies. Frequently, the initiative launch order will appear very front loaded—as if most of the opportunities are to be launched in the current month. Obviously, this is impractical. Before the order resembles a feasible plan, it must account for the resource requirements for the initiatives.

The next step in ordering the initiatives is to match initiative resource requirements with available resources (i.e., resource allocation). This entails a bit of guesswork. The actual point in time when the resources are needed may not be known until the initiative is under way. In lieu of this timing, assume that the resources are needed when the initiative kicks off. As information becomes available, this assumption can be adjusted. Returning to the initiative chart, the initiatives were initially ordered based solely on prioritization criteria. Then we incorporated dependencies into the order. Now we are going to assign limited resources to the initiatives. As we progress through the resource-allocation process, as long the order is maintained, the initial prioritization carries through the remainder of the allocation process.

Limiting resources should have been identified during the scoping of the initiative. At this point, we need to determine the availability of the limiting resources. Do not limit this review to solely tangible assets (i.e., money, machinery, facilities, materials, etc.) but also include human resources—especially needs for expertise or specific process owners. If a limited resource is not immediately available, identify the point in time when it will be available. A resource chart such as Table 7.4 is helpful to track resource utilization.

In general, the use of limited resources and cash requirements can be managed in weekly or monthly increments. Sometimes weekly is too detailed; other times it is not enough. Experience and iteration will aid in determining the appropriate timing periods for scheduling.

Once the limited resources and their availability are known, the scheduling process is fairly straightforward. Begin with the first initiative, and examine its resource requirements. If the resources are available, theoretically, the initiative is ready to go. Assign the resource to an initiative using a utilization table similar to Table 7.4. The first initiative retains its rank as the first initiative to be launched (assuming that it has not already started). Proceed to the second initiative. Review its resource requirements, and determine their availability.

TABLE 7.4   Limited-Resource Tracking Chart

| | Resource Needs (Project Experts) | | | | | |
| | Jan | | Feb | | Mar | |
Initiative	PM	Lean	PM	Lean	PM	Lean
Working Capital Initiative	1	0	1	0	1	0
Sales Force Redesign	1	0	1	0	0	0
New Product Launch	0	0	1	1	1	2
IT Efficiency Study	0	0	1	2	1	2
Needed Resources	2	0	4	3	3	4
Available	3	2	3	2	3	2
Variance	1	2	−1	−1	0	−2

Schedule this initiative to start when the appropriate resources are available. Once again, reserve the needed resources, and move on to the next initiative. Continue evaluating the full portfolio of initiatives in this manner. Make the starting point for each initiative as early as possible after the resources are available and dependencies are met. This may mean that because of the unavailability of some resources, a less beneficial initiative may be slated to start before initiatives with greater anticipated returns.

There is also an opportunity to buy specific resources that are not immediately available in order to remove resource constraints. One frequently purchased asset is outside expertise, such as management consultants or other specialized professionals. But when acquiring limiting resources to mitigate shortages, the additional cost needs to be factored into the business case for the specific initiative. This additional cost may push the initiative back in the prioritization order, or it may have minimal impact. The question is whether the tradeoff for time is worth the cost.

When building the schedule, there is some wiggle room in determining the exact order. The question always arises as to how to account for initiatives that are already in process—where resources were allocated, teams assembled, and work is underway. On the initiative chart, it helps to identify in-process initiatives because they deserve unique treatment. Stopping and restarting initiatives is inefficient. When work is halted, team members may transition into other roles—requiring new team members to be assigned and brought up to speed. Knowledge and experience gained may be lost. In short, a restart is not just picking up and moving forward—it is a refresh. In these situations, it is a judgment call. An in-process initiative generally should receive a healthy dose of favoritism and be given the green light over other initiatives that are forecasted to deliver a comparable benefit. The obvious exception is when an initiative is misaligned with the enterprise's strategic direction or is no longer expected to deliver value. Such initiatives are candidates for elimination altogether.

One additional note when building the initiative order: most enterprises struggle to execute more than a handful of major initiatives at the same time. Enterprises just do not have the focus or the breadth of resources to tackle many large initiatives simultaneously. After reviewing the initiative order, the leadership council may reorder the largest initiatives to limit the amount of major change during any single period.

The final step in building the launch schedule for initiatives is a reexamination to identify collaboration opportunities. Often there is a benefit to executing two initiatives in tandem. The benefit may be in collectively designing the solution to ensure that it is appropriate for both initiatives, or there may be an efficiency to be gained by coordinating information gathering or managing the eventual release of solutions. Review the initiative chart to identify collaboration opportunities. When it makes sense, the schedule can be adjusted to take advantage of coordinated efforts. Other times, heightened communication between the two initiative owners may be sufficient.

Although many individuals participate in the process, I strongly recommend that a final review of the portfolio be completed on a monthly or quarterly basis. This review delivers two benefits. First, it allows for buy-in as to the prioritization and respective commitments of the enterprise team. And more specifically, it confirms that the process sponsors and initiative owners accept the business cases and content of the specific initiatives that are assigned to them for execution. Second, a review allows a moment of introspection as to the enterprise's priorities. Although the ranking aligns the launch order to the prioritization factors, strategic planning is an intuitive act; the assumptions behind the strategic initiatives may change and require adjustments to the schedule before the innovation plan is launched. Strategic review is an ongoing activity sometimes requiring changes on the fly.

With approval of the launch order, the innovation plan is ready to be executed. Teams are formed for the top initiatives. Resources are allocated or procured. Communications are launched.

MAXIMIZING THE VALUE OF THE INNOVATION PORTFOLIO

Theoretically, every initiative delivers a contribution—the overage of benefits delivered versus costs accrued and discounted to reflect the relative value with the passage of time. Therefore, the theoretical net benefit of an enterprise's portfolio of initiatives is the sum of the net present value of each initiative. Because the value of an initiative depends on the timing of the costs and benefits, the value of the overall portfolio of initiatives depends on the order in which the initiatives are executed. From a portfolio-management perspective, this means that the value of the portfolio of initiatives fluctuates based on the start, finish, and duration of any initiative.

If the initiatives generating the most value are pushed behind lower-value initiatives, the value of the portfolio falls (assuming a normal business environment). To maximize the value of the portfolio (assuming that value generation is the primary prioritization criterion), initiatives delivering the greatest value need to be moved to the front of the line. In fact, the portfolio that orders initiatives by their net benefit (all else being equal) maximizes the value delivered by the innovation portfolio.

ONGOING MANAGEMENT OF THE PORTFOLIO

The prevailing business practice today is that after an initiative is launched, the focus swings from a move-forward decision to completing the initiative as expeditiously and at as low a cost as possible.

Deviations from the original plan are identified and then rectified through the approval of change orders and budgetary adjustments. Only in the most rare instances are initiatives reexamined to confirm their validity.

Across corporate America, multitudes of initiatives move forward despite the arrival of circumstances that reduce or negate their anticipated benefits. Project teams spend time and money developing solutions that will never be implemented. Why? Sometimes information does not flow to the right people who make the decisions. Other times the assumptions on which an initiative is predicated are never documented or understood. And then a considerable number of times the initiative keeps getting funded because of politics. This last failure is perhaps the most alarming.

As individuals climb the corporate ladder, they adopt the belief that advancing their careers requires the avoidance of any blemish on their record. A major epidemic spreading across leadership teams is the fear of failure. Leaders know that any real or perceived failure may well become a topic in promotion discussions. Closing down a corporately blessed initiative, especially a major one, is perceived as a leadership failure on a colossal scale. "He or she couldn't make that one work. Why would it be any different if he or she were in this role?" This is an often-stated justification for pigeonholing an individual. Not surprisingly, leaders avoid making the call to shutter initiatives. In the current leadership mindset, it is far better to let an ineffective initiative continue than to admit failure.

For example, a marketing executive at a Fortune 300 retailer hired a software vendor to deliver a workflow solution. The selection was a poor choice, based on a relationship instead of the tool's capabilities. The project was budgeted for six months, but two years later it was far from complete. Off the record, the vendor apologetically stated that the work was beyond their capabilities. The project team identified alternative solutions and presented them to the senior marketing executive. To their amazement, the marketing executive rebuffed their investigation and instituted a gag rule—eliminating any further consideration of alternatives and declaring that the organization would "land the plane with the group they took off with." In hard costs, several million dollars were tossed down the drain. When questioned about his decision, the executive admitted that he did not want to be stigmatized with a failure.

In an enterprise with a robust portfolio-management function, the leadership council regularly reviews initiatives and their base assumptions. As the conditions on which the initiatives are based evolve, the business cases and resource requirements for the initiatives change as well. Initiatives originally estimated to provide major benefits now may produce minimal or no value. Depending on the magnitude of the change, the portfolio may require the creation of a new initiative, the abandonment of an existing initiative, or simply the assimilation of new resource requirements into the initiative. This fluidity of information allows resources and attention to be diverted to more promising endeavors. Because the innovation plan is collectively developed and managed by the leadership team, blame does not stick to any individual. This is a huge (yet inadequately recognized) benefit of the collective management of the innovation plan—the ability to minimize political footballs and sacred cows, focus on beneficial initiatives, and eliminate wasteful investments. This active review and management makes the innovation plan more dynamic than static. Thus, although an innovation portfolio works for enterprises tied to an annual strategic planning cycle, the companies that adopt dynamic strategic planning enjoy a major competitive advantage—the ability to react to market forces faster than the competition.

The ongoing management of the innovation portfolio is the responsibility of the leadership council and overall process-governance organization. Management of the innovation plan is akin to running an enterprise-wide program-management office (EPMO), but the responsibilities transcend that of a typical EPMO. Unlike an EPMO, which primarily tracks the status of work efforts, a portfolio-management team monitors the rationale behind the initiatives and reprioritizes them as circumstances change. This brings us to the responsibilities of the leadership council in regard to managing the portfolio of initiatives. To recap, the primary responsibilities of the leadership council are as follows:

- Select prioritization criteria, and rank them according to enterprise priorities.
- Build teams to scope and develop business cases for initiatives.
- Confirm the innovation plan at regular intervals.
- Obtain and allocate resources to support the innovation plan.
- Continually assess the competitive positioning of the enterprise.
- Monitor the ongoing innovation plan.
- Collect new or adjusted process requirements from internal and external feedback loops.
- Reevaluate the assumptions used to build the initiatives.
- Adjust the resource allocation for the initiatives.
- Ensure that coordination and collaboration occurs between initiatives.
- Address any issues or risks brought to the leadership council.
- Identify and make adjustments when the situation warrants such action:
  - ▲ Evaluate new initiatives for inclusion in the innovation portfolio.
  - ▲ Abandon ongoing initiatives no longer relevant because of current developments.
  - ▲ Reprioritize the portfolio when the prioritization criteria change.

▲ Assimilate new requirements into existing initiatives if the current initiative trajectory is incorrect based on new information.

▲ Understand enterprise-level change impacts to business areas/customers, and adjust accordingly.

BENEFITS OF AN INNOVATION PLAN

Using an innovation plan as the means to methodically innovate elevates an enterprise's ability to implement improvements to operations and enhance its product offerings. Clarity around the actual improvements and widespread understanding of those changes are the immediate benefits. No longer do leaders and managers need to debate, question, or theorize as to the true intent of an initiative. Plans and priorities are embedded in the innovation plan and defined in the language of process. For enterprises embracing a process-based philosophy, the advantages are significant. The innovation plan

- Creates a customer-focused culture.
- Provides clarity of strategic intentions of executives/strategic planners.
- Focuses on the concrete and not the abstract. Bases adjustments on processes—the foundation for value creation in every enterprise.
- Ensures alignment of processes (primary and supporting) with strategic and operational initiatives.
- Uses a holistic view of the enterprise to ensure that change is introduced on a scale and in a coordinated manner that optimize the value gained by improvements.
- Avoids or minimizes the risk initiatives that improve localized areas at the expense of the overall system.

- Provides an efficient framework to disperse resources, money, and managerial focus in alignment with the prioritization. Allocates resources based on actual need—not just an addition to the prior year's budget.
- Reduces the need for continual communication about the intent and scope of an initiative. Because of the relative top-down nature of initiative creation, coordination with other stakeholders is identified early in the process and built into the team structure.
- Minimizes managerial turf wars that plague operations and encumber the performance of improvements by getting all leaders, supervisors, and managers on the same page as to the focus of the enterprise.

With completion of the work to build the innovation plan, the leadership is pulled together for a final confirmation of the innovation plan. Their vote for the allocation of resources signals both an organizational commitment to the innovation plan and a mandate for action. Although it sounds easy enough, contemporary enterprises are tied to their existing structures and processes that have been inherited—over decades in some instances. To innovate is to break the chains of yesterday and rebuild structures, roles, and processes. Implementing a process-based approach is easier said than done—but the benefits are worth it.

8

Implementation of a Process-Based Approach

Every enterprise conducts business using a set of practices expressed through processes, structures, policies, and systems. Behind these practices is an equally important collection of operating practices that are unspoken and unwritten. This second set is rooted in the traditions, biases, practices, tenets, and habits that propagate throughout the human interactions inside the enterprise. And it is this set that defines an enterprise's culture—its very personality and how discussions are conducted, how decisions are made, and how the organization mobilizes resources to chase opportunities. The widespread acceptance of these operational practices acts as a stabilizer for decision making—a common lens for framing choices. But these same practices come at a cost. As the competitive environment shifts, these beliefs and policies limit the consideration of alternative courses of action. They anchor examinations of choices to what is perceived as proven, tested, and safe. And unfortunately, these limitations can spell doom when situations require major shifts in strategic direction to get the enterprise in a competitively viable position. What worked in the past feels safe to leadership teams. Stepping outside the boundaries of the existing conventions invites risk—or so the thinking goes. An aversion to disruption and big change pervades today's leadership ranks. Iconoclasts and free thinkers are viewed as troublemakers who need to soak in the culture before speaking their mind.

To a large extent, these limitations are heavily affected by the size and organizational structure of the enterprise. In stark contrast to the behemoths of the corporate world, entrepreneurial ventures are comparatively nimble. Because of limited resources, the leadership team often wears many hats and can be found actively engaging with customers on the front line. This proximity begets a ground-level understanding of the customer and a comparatively quick identification of issues and market opportunities. Armed with both knowledge of the situation and the authority to get things done, entrepreneurial leaders are well positioned to realign resources and nab the advantage.

From my consulting experience, I learned that when outright asked, most managers will respond that they and their departments are adept at changing. From their perspective, their worlds are in constant flux as new leaders arrive and promote their way of doing business. But the change these managers refer to is more akin to settling in and returning to a business-as-usual environment after the latest restructuring. When a new leadership administration takes over, or when performance drops below leadership's expectations, senior leaders frequently resort to a corporate reorganization. Today's reorganizations are primarily a shuffling of the management team. Several unfortunate managers are let go, but most are simply reassigned to new roles and responsibilities. Although there may be new names in dozens of positions, in most cases these "reorgs" lack any true fundamental change to the way work is designed and performed. The average worker punches his or her time card and follows the same patterns he or she would on any other day. It is a fresh coat of paint on the corporate facade, but little more. Once the reorganization is complete, the same work activities equate to the same results. Over time, the lack of progress from these reorganizations breeds lethargy and tacit resistance among the workforce. People quit believing that any transformational effort will really change anything. The prevailing sentiment is to just wait around, and things will revert to normal.

WHY ESTABLISHED ENTERPRISES STRUGGLE TO ADAPT TO A CHANGING ENVIRONMENT

For a moment, think about the railroads of yesteryear, where icons such as Carnegie, Gould, Pullman, and Vanderbilt amassed immense fortunes. In their heyday, the railroads consisted of a network of rails that traversed the land—spanning rivers, burrowing under mountains, connecting cities across the United States, and serving as the primary means of long-distance transport for generations of travelers. From their vantage point atop the transportation industry, the railroads attained seemingly insurmountable strategic advantages. They understood business and leisure travel to an extent never previously envisioned. They accumulated vast stockpiles of cash to invest. In fact, if they played their cards right, they were on their way to reign over the transportation industry for decades. But their fortunes unraveled as the world changed. A wave of inventions including the automobile and airplane enticed customers with faster and more convenient alternatives. The railroad magnates took their eye off the ball and failed to adapt. These days, although railways span more miles in the United States than in any other country in the world, railroad traffic is a minor percentage of total transportation mileage.

Where in 1900 there were 132 type 1 railroad companies (revenues over $350 million adjusted), there are only 7 today. Amtrak is the sole intercity passenger railroad. Although once hailed as crucial to America's fortunes, the performance of the remaining seven railroads is mediocre in their best years—and horrific more often than not. Despite their significant competitive strengths, the railroad companies failed to keep pace with advancements in transportation. Why did they fail on such a grand scale?

With the turn of the twentieth century, the United States was rich with resources and the spirit of innovation. Automobile companies such as the Olds Motor Vehicle Company in 1902 and the Ford Motor Company in 1903 began selling cars to the wealthy and elite. Then, in 1908, the Model T was manufactured on Ford's assembly lines, and the automobile was produced at a cost that made it within the means of the masses. Air travel also was taking off. With the conclusion of World War I, a mass of aviators returning from the battlefront spurred the postal service to explore mail distribution by air. After an initial failure, a network of contractors was hired to fly mail to destinations across the United States. From these initial 12 contractors, companies such as Delta, America, United, TWA, and Eastern Airlines were launched. Shortly thereafter, commercial air travel took off. While the transportation industry was rapidly changing, the railroads stuck to their business models and practices. Although well capitalized and armed with superior knowledge of passenger and freight travel, they opted not to engage in the automobile or airline industries and missed a gargantuan opportunity to build a comprehensive transportation network serving customer and freight needs on a grandiose scale. But before you scoff at the failure of the U.S. railroad companies, recognize that they are not atypical in their shortsightedness. Examples abound of companies, industries, and governments that reigned for decades—even centuries—before the winds of change pushed them to the pages of history texts.

Making Big Change Happen

From my observations, it is an undeniable truth that most enterprises struggle mightily when confronted with change of any significant level. The culture and institutional practices born of historical success have effectively vaccinated leadership teams against future innovation considerations. Adjusting strategy or making profound market adjustments is taken off the table in many boardroom discussions. The personal ambitions of leaders conflicts with the risks of sailing into unchartered areas. Having seen other leaders get the boot for failures, they know that such a blemish on their record may well derail their career and affect their personal livelihood. Such a mentality is epitomized in popular platitudes such as "If it ain't broke, don't fix it." Rationalizing their inaction through public statements that things will be just fine, leaders conveniently ignore the subtle and not-so-subtle shifts in the markets. But the clock continues to tick—and eventually the reality sets in that the enterprise is losing ground. Something must be done.

Today's enterprises are not built for periodically reinventing the way they do business. Market and other insightful information flows poorly from the front lines to the decision makers in the enterprise. Plans take months to create. Getting buy-in and coordinating execution takes even longer. Although there is the occasional exception, enterprises rarely seek to reinvent themselves and truly change their operational practices in but a handful of circumstances. The first is when a new senior leader takes over and seeks to put his or her imprint on the organization. When Steve Jobs returned to Apple for the second time, he reinvented the company's core operational practices. He cut the company's product offerings, eliminated a good portion of the distribution channels, and set out to develop a pipeline of new, innovative products. This case is also equally illustrative of another instance when enterprises undertake large-scale reinventions—when their backs are to the wall. When the threat is so great that the doors may close unless drastic action is undertaken, the leadership team may galvanize around a fix that fundamentally alters the enterprise's way of doing business. But even when the odds are firmly against them, many leadership teams will stand firm to their accepted business practices and ride the corporate ship as it descends beneath the waves. This leaves us with a final instance when enterprises drastically change course—when an undeniable success occurs because of a substantial shift in an enterprise's operational practices. For example, think of when Henry Ford institutionalized the assembly line. If the competition hoped to compete against the Ford Motor Company, adopting the assembly line was an imperative.

Regardless of the catalyst that precipitated the action, all the aforementioned enterprise transformations were possible because some visionary individual convinced (or was) a senior leader with the authority to get the full enterprise united behind the program. This type of visionary is a unique individual—possessing the awareness to see that things are not operating as well as they could and the conviction to promote the potential of a new idea. These champions have that rare capacity to connect the dots and see beyond the chaotic and dysfunctional environment in which they currently reside. To make the change happen, these champions must build both awareness and momentum.

Adopting the practices and structures of a process-focused enterprise is definitely classified as large scale—on a level exceeding the modern-day reorganization. In fact, it drastically parts with the traditional manner of organizing people and establishes new practices and structures to manage the enterprise. The magnitude of these changes makes it impractical to jump forward using an "all-in, full-steam-ahead" approach. In lieu of a "big bang" approach, a more gradual, concerted, and systematic approach yields the greatest probability of success. My recommended approach is to make the big change appear minimal—to make it palatable until that point of no return when the process-based approach is delivering an undeniable advantage. This is the aim of the visionary—to ostensibly deliver incremental value until senior leaders cannot help but be curious about the next steps. At that point, however, the sell and plan need to be buttoned up and ready for presentation.

Start with a Small Win

One of the great strengths of a process-based approach is its ability to define work activities and provide an enriched view of business operations. One of the simplest ways to build awareness and momentum is to implement the principles of process management in a part of the enterprise. This includes mapping the processes, setting process goals, and gradually building a plan for innovating the area using process adjustments as the language of change. Once things get moving, the transparency of intent and ease of interaction with this team will capture attention. The team becomes easy to do business with. As this occurs, the visionary catalyst needs to have the story ready to go—a picture of how the enterprise could operate if everyone used a process-based approach.

Build a Convincing Story

Before making the sell, take the time to clearly note what the transition will mean on an enterprise level. A process-focused enterprise is a new animal to most leaders. Educating them on the difference between a process-focused enterprise and the more traditional functionally based enterprise will be an eye-opening experience.

The sell not only identifies the benefits to be gained but also spells out how the practices of managing and innovating the enterprise will change for the leadership and managers. A short list of benefits might include

- *Customer connectivity.* A process-based approach revolves around providing superior products and service for the customer and recalibrating the enterprise to support these customer-specific aims.
- *Clarity and shared awareness.* This involves a universal and crystalline understanding of the operations and capabilities of the enterprise and immediate access to know who does what.
- *Innovation foundation.* This is a foundational view of the enterprise on which to base improvement activities.
- *Effective strategic planning.* This is the forging of a simplified and concrete link between strategy and its execution.
- *Efficiency and cost reduction.* This involves building a responsive structure to eliminate excess and focus resources on the key activities to improve performance and accelerate capability building.

Although the benefits paint a picture of what could be possible, it is equally important to bring the idea home. Communicate the benefits in the enterprise's terms. What will it mean for operational challenges existing today? How might interactions with customers be different? What have similar enterprises done to achieve success? Why not us?

Deliver a Comprehensive Plan

A frequent response from leaders after being introduced to a process-based approach is to argue, "We already have many of the elements of this approach." This statement is equivalent to, "We have some elements of a bridge." A working process system depends on all the components working in tandem to make innovation effective and efficient. Only having parts of the solution means that many advantages of the approach remain hidden in the closet.

When asked to provide input on the transition, change agents need that elevator speech—several simple statements as to the anticipated benefit of the approach. Put it in simple terms. For example, "What would a group of executives say if they were told that they could have a deeper and richer understanding of their business and a vastly increased connectivity between their strategy and daily execution—and have this capability delivered in less time than it takes to develop the corporate strategy today?"

A Comprehensive Plan for Transitioning to a Process-Focused Enterprise

In general (although there are exceptions), most leadership teams lack a relatively accurate view of their customers, the ground-level operations of their enterprises, or exactly where to focus resources to take advantage of opportunities. At this stage, the plan is to tackle these items in a relatively unobtrusive manner—and thus take the first steps on the journey. For these business challenges, a process-based approach delivers immediate benefits—and it does so without shaking too many trees.

The recommended transition consists of three phases: start "lite," build momentum, and formalize and institutionalize (Figure 8.1). Each phase expands the transition but does so under the guise of simplifying the way work is completed while building new awareness of the current operational processes as well as the external forces affecting the enterprise.

FIGURE **8.1** Three phases of a transition to a process-based approach.

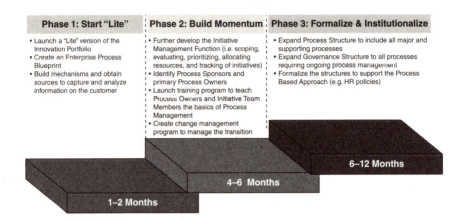

Phase 1: Start "Lite"

- Launch a "Lite" version of the Innovation Portfolio
- Create an Enterprise Process Blueprint
- Build mechanisms and obtain sources to capture and analyze information on the customer

Phase 2: Build Momentum

- Further develop the Initiative Management Function (i.e. scoping, evaluating, prioritizing, allocating resources, and tracking of initiatives)
- Identify Process Sponsors and primary Process Owners
- Launch training program to teach Process Owners and Initiative Team Members the basics of Process Management
- Create change management program to manage the transition

Phase 3: Formalize & Institutionalize

- Expand Process Structure to include all major and supporting processes
- Expand Governance Structure to all processes requiring ongoing process management
- Formalize the structures to support the Process Based Approach (e.g. HR policies)

1–2 Months

4–6 Months

6–12 Months

Phase 1: Innovation Portfolio, Enterprise Process Blueprint, Customer Information Gathering

The aim of phase 1 is to introduce stability (including common terms/views/reports) and order into the management of ongoing and planned innovations. Again, the intent is to deliver incremental value by injecting clarity and precision into the innovation cycle. To achieve these objectives, the initial focus is to launch a "lite" version of the innovation plan. By introducing a high-level system to manage initiatives and allocate resources, the leadership team immediately gains awareness of ongoing innovation activities. More often than not, this initial foray into managing innovation usually points out glaring deficiencies in current business practices.

A "Lite" Innovation Portfolio

When an enterprise is undergoing a transition to systematize the planning and execution of initiatives, it becomes necessary to set boundaries on the scope and nature of those initiatives in the portfolio. Commonly, such guidelines revolve around the size of the initiative and whether it is cross-functional or not. The initiatives fitting the criteria become the working innovation portfolio. However, what sounds like a simple exercise can often be a major undertaking. At this point in time, the initiatives are rarely developed using a rigorous process to clearly define the end goal, the business case, or the resources required to bring them to fruition.

But excessive analysis at this point can paralyze the innovation process. For this reason, take a shortcut and use high, medium, and low ratings to rank the net benefit delivered, timeline for completion, or other prioritization factors when data is not readily available. Using this "lite" approach, the highest-priority initiatives can be pushed forward before the full-scale deployment of a robust initiative evaluation process. Time is money in today's hypercompetitive markets.

Creating an Enterprise Process Blueprint

While the "lite" innovation plan is being implemented, a team should be commissioned to build the initial enterprise process blueprint. A complete approach for documenting an enterprise process blueprint was presented in Chapter 5. The blueprint is invaluable to leaders attempting to grow awareness of the operational capabilities of an enterprise. When first created, an enterprise process blueprint is an eye opener—equivalent to shining a flashlight into the recesses of the operational structure where unclear ownership, ambiguity, and misunderstandings flourish. As this initial foray into mapping the enterprise's processes is completed, it invariably sparks a collective review and epiphany about the current operations.

The blueprint will likely be messy at this point—with "bolt-on functions" and other inefficiencies immediately visible. Heritage and non-value-added processes abound. This is a result of the degree to which an enterprise is functionally focused—and quickly conveys the need to focus on the customer and reposition the enterprise to serve the customer.

The blueprint is the launching point to understand the focus (or lack of focus) that exists in the current state of the operational structure. It serves a holistic view of the enterprise and allows for clarity around the capabilities of the enterprise and where platforms exist for new offerings. The act of building the blueprint helps identify functions/processes where ownership is confusing or even missing and it leads to insightful questions:

- To what degree is customer focus embedded in the enterprise? Where are the touch points with the customer? How are they managed? Are there customer feedback loops to leadership and other appropriate individuals?
- Where are the work handoffs across functional borders, and how are they affecting the performance of the process? Are there bottlenecks preventing scalability of major processes?
- Does work flow via a simple path through the enterprise? Or does confusion and chaos reign in certain areas?
- Is work completed in the correct place and by the correct performer? Could the work be transitioned to other performers or locations better equipped to complete the work?
- Are there duplicative functions that could be consolidated in designated centers of excellence to aggregate work and resources to enhance efficiency? (In the case of a conglomerate or diversified company, obtaining this perspective may require a review of multiple enterprise process blueprints.)

Although this new vantage point may bring issues to the surface that initially frustrate and confuse leadership teams, it quickly becomes foundational to conversations regarding the improvement of operational capabilities—and with an unprecedented clarity. Although it is a very powerful tool, the enterprise process blueprint is only an initial step to grow awareness of the enterprise's operations. Equally important in this initial phase is the development of a richer understanding of the customer. "The first step in exceeding your customer's expectations is to know those expectations."[1]

Grow Feedback Channels and Awareness of the Customer

As spelled out in Chapter 2, the first facet of an innovative enterprise is for it to be customer focused. Although some amount of customer research exists in nearly every enterprise, the data rarely provides sufficient detail to plan future innovations. A process-focused enterprise uses customer data as the fuel for plotting adjustments in the process structure—and delivering superior customer experiences. Achieving this state entails building feedback loops from the frontline employees who interact with customers on a daily basis and merging this information with that from other sources, including customer analytics, research studies, and qualitative customer analysis sources. Because of the importance of the customer, this is always an area where improvements can be made. Some of the immediate opportunities for improvement include identification and management of customer connection points, capture of customer data from frontline workers, and compilation of this information into a format conducive to distribution. Techniques and approaches to gather and synthesize customer information were presented in Chapters 2 and 4.

At this juncture, the enterprise has taken several significant steps on the path to a full process-based approach. Whereas implementation of a "lite" innovation plan, creation of an enterprise process blueprint, and development of customer feedback channels are relatively benign and unobtrusive adjustments, they put structure where previously there was none. In most instances, these changes breed minimal resistance. As the three pieces come together, opportunities

to improve the enterprise surface. One surefire way to build awareness with leaders is to review the current-state enterprise process
blueprint while overlaying the customer analysis. This approach
never fails to bring a number of commonsense initiatives to the surface, and it eases the transition to phase 2.

Phase 2: Kicking Off New Initiatives and Stabilizing the Portfolio-Management Function

Phase 2 builds on the foundation put into place during the initial
phase and pushes the management team to amble further down the
trail to a process-based approach. Even more important, it builds
knowledge and processes that are foundational to successfully driving future innovation. Depending on the degree of attention and
interest received, phase 2 may be the appropriate time to get a team
up and running to guide the entire enterprise through the transition. And this precipitates other activity as well, including formation
of an enterprise resource pool to staff initiatives, development of a
financial planning group to oversee the initiative-development process, and setting up basic process training classes and other related
activities.

Enhance the Initiative-Management Processes

With an innovation portfolio loosely in place, the next step is to
increase the robustness of the processes used to evaluate and prioritize initiatives. An important accomplishment in this phase is
translation of the enterprise's priorities into a comprehensive portfolio-management function that evaluates, ranks, and orders improvement initiatives. Embedding the priorities of the leadership team into
the initiative-development process effectively institutionalizes the
team's relative significance to the future prosperity of the enterprise.
This activity requires leaders to commit to the innovation plan and
the decisions it creates. The shortcut is to use standard prioritization

factors, including net present value created by an initiative, initiative dependencies, and resource requirements.

With the prioritization decision made, the tools and techniques to analyze and design improvement initiatives need to be further developed and embedded into the planning routines. Because these processes are generally underserved and underdeveloped, a review of the full gamut of initiatives is necessary to ensure that the prioritization is a fair and accurate ordering based on the confirmed priorities. Initially, the executive committee or other governing council that makes the major strategic decisions serves as the ad hoc leadership council. As new initiatives are released for evaluation, the leadership team performs the portfolio-management responsibilities presented in Chapter 7. Simply meeting on a regular basis to review the initiatives formalizes and legitimizes the innovation plan. There will be speed bumps, but now the leadership knows how its forces are arrayed and where activity is taking place. Almost immediately, leaders and employees begin to grasp the rationale behind the portfolio-management process and use it as the framework to explore new opportunities.

Select De Facto Process Owners and Sponsors to Shepherd Initiative Solutions

As the initiative-development process begins to hum, it is a perfect time to introduce the role of process facilitator. In most large-scale enterprise transitions, a core team is engaged to build the guidelines and tools to support the new portfolio-management activities. The individuals fulfilling this role gain a ground-level understanding of the innovation cycle and a thorough education on the process-based approach. With this background, they are excellent candidates to be process facilitators. At this phase of the transformation, process facilitators are the smoke jumpers in the enterprise. When confusion reigns or progress is stymied, they leap into the fray and use their knowledge and skills to make progress and tamp down the fire.

They may serve as trainers, leaders, facilitators, and coaches during this period. As the evolution proceeds, gradually their responsibilities settle into the more traditional role of process facilitator.

With the engine running, the focus shifts to addressing shortcomings of initiative management, tailoring the prioritization to the enterprise's priorities, and institutionalizing the knowledge to manage process-based initiatives. As initiatives are pushed into the chute for execution, planning of the end solution needs to incorporate the eventual handoff to a team that performs the process in the normal course of business. Even at this early stage, it is beneficial to begin identifying de facto process owners who will supervise the process once it is up and running. The functional department heads are convenient candidates to groom into these roles because they already act as the subject-matter experts for their areas. As more initiatives are launched, a governance structure begins to take shape as more de facto process owners are engaged.

Institute Process Training

The final part of phase 2 is the launch of a training program to push the concepts of process improvement to a larger audience. Because the level of knowledge differs based on the role an individual plays, any training curriculum needs to cover the breadth of the roles and responsibilities of the full process-governance organization. Although process-improvement skills are foundational, the initial courses ideally focus on helping individuals to navigate the major components of a process-focused enterprise. As a sample curriculum, the initial courses might be as follows:

- Course 1: The Enterprise Process Structure and Process Management (for leaders and process owners)
- Course 2: The Innovation Portfolio (for leaders and process owners)

■ Course 3: Initiative Execution (for process owners and initiative stakeholders)

These courses suffice to make the early adopters aware of the basic constructs of a process-based approach. As the principles of process management are adopted, additional coursework can expand on the basics communicated in these initial classes, and eventually, the training will expand beyond just the initiative teams to include leaders, managers, and workers throughout the enterprise. Fully capturing the benefits of a process-based approach requires the full employee population to be trained on processes—giving them the ability to weed out heritage processes and continually improve their areas of the enterprise. The goal is to incite the workforce to aggressively identify improvement opportunities and simultaneously provide the framework to evaluate those opportunities and begin leveraging the full resources of the enterprise in the most expeditious and beneficial manner.

At this point, the enterprise has adopted (occasionally unknowingly) many of the practices of a process-based approach. Although the only formal structure that exists is the portfolio-management function, around the enterprise individuals are beginning to catch onto the new way of thinking and organizing their work efforts. An enterprise can operate in this mode for an extended period—at least until any resistance subsides and associates become comfortable with the new approach.

Phase 3: Full Steam Ahead (Building Out Process Structure and Governance Model)

As initiatives are launched and completed, a portion of the process structure is rebuilt, and other sections receive at least token improvement treatment. Yet there are most assuredly gaps—areas outside the sphere of improvement efforts where heritage and poorly designed processes still operate, areas that would definitely benefit from attention. Over the years, I found these back-burner areas to be some of the most fertile areas for investigation. Indeed, teams focused on delivering immediate cost savings are advised to closely inspect these areas because they are rife with duplication, inefficiency, and an abundance of opportunity.

Finish Documenting the Process Structure

Mapping these back-burner areas will take an extended period of time if left to a convenient time, and the potential benefit will be pushed into the future. Why eliminate from consideration what might possibly be some of the biggest opportunities? The smart approach is to map these unchartered areas and assign process owners to begin examining them for opportunities. Knocking the mapping out in a timely fashion frequently requires a dedicated team. If they are available, process facilitators are the ideal resources to complete this work. However, if they are knee deep in managing initiatives and getting the leadership involved in the portfolio-management function, using them may delay the overall transformation. The alternative is to use process owners or other experienced process performers to complete this work. If appropriately qualified resources are not available, hire an external team to complete this work. From my perspective, the best option may be to hire experienced process consultants and pair them with promising internal resources. In this manner, the consultants' experience is leveraged to get the work done while simultaneously the internal resources gain an education. Locating consultants with strong process skills is rarely a problem because this type of work is their bread and butter.

As mentioned in Chapter 5, not every process requires mapping or, even more important, active management. There are a number of processes in every enterprise that are not suitable for mapping because of the variability of their inputs, outputs, or the process itself. Customer service and creative processes fall into this bucket. For these processes, procedures or guidelines suffice for managing their performance to desired enterprise standards. The degree of active management is significantly lower than with other processes. And there are always processes that are executed so infrequently as to be immaterial and require no management. The juice is just not worth the squeeze. As for general guidelines, a process should be assigned a process owner for active management if it meets any of the following requirements:

- The process is repeatedly and consistently executed in some interval of time.
- The process provides a valued output.
- The process consumes significant resources.
- The process is foundational in that it provides important capabilities to the enterprise.

In other words, which processes would deliver even greater benefits if a process owner were continually engaged to monitor its performance and make improvements? Using these guidelines, the remainder of the process structure should be built using the techniques presented in Chapter 5.

Formally Assign Process Owners and Sponsors

Simply defined, *process governance* is assigning managers to own parts of the process structure. Although the relationship between the two is symbiotic, the creation of a governance structure is entirely dependent on the existence of the process structure. As the processes and their connectivity are brought to light, process sponsors are appointed by the senior leadership team to oversee groups of similar processes. In most enterprises, the individuals most qualified to immediately occupy these roles are currently serving in traditional leadership roles (e.g., senior vice presidents or vice presidents). In many instances, the ideal candidate does not exist. Compromises are necessary until recruiting and training programs are in place to find and develop the ideal candidate. To get folks into roles, individuals may be chosen based on their leadership credentials, although they lack process acumen. Over time, these individuals will grow into the role or a more suitable candidate will be found. As with any major transformation, not everyone who started on the journey will arrive at the end.

As process sponsors are named, they are, in turn, responsible for filling open process owner roles in their areas. At this early point, a candidate's familiarity with the subject process takes precedence over his or her process expertise. Process skills can be bought or rented when needed—but operational knowledge is only available internally.

Eventually the process structure is fully built, and names are attached to the boxes. However, it will take time before the enterprise is hitting on all cylinders. It simply requires a number of cycles before all the sponsors and owners possess the knowledge to be effective in the new environment. Many enterprises attempt to shorten the learning curve by sponsoring knowledge-sharing sessions across levels (e.g., process sponsors, process owners, process facilitators, and initiative-team members). These sessions serve multiple purposes. First, individual process owners have a forum to discuss their daily challenges and, when appropriate, escalate common challenges to upper-level leadership. Second, these meetings are useful to build awareness of the endeavors of others at a similar level in the enterprise and to plan collaboration opportunities. And, of course, there is the additional benefit of providing a forum for individuals to network internally. Such a meeting is an excellent way for process owners (and sponsors) to develop awareness of candidates for future open positions.

With the creation of the process-governance structure and the initial innovation plan, the company is well on its way to becoming a process-focused enterprise. An initial goal of the acting leadership council is to formally approve the first slate of initiatives to be the initial innovation plan and to assign the resources to make it happen. From this point on, it is business as usual. The leadership council functions as a super program-management office—maintaining a continual focus on the customer, monitoring all facets of operations, and deciding on how to take advantage of upcoming opportunities. In every enterprise, though, the real work takes place in the initiatives—the place where the rubber meets the road and where innovation succeeds or fails.

9

Executing a Process-Based Initiative

" **C** hanging the engine while the airplane is in flight" is how one consultant described a massive reengineering effort. An ever-changing world forces enterprises to adapt or die—and to make those adaptations while simultaneously keeping operations humming. Many enterprises expect their employees to play dual roles—to design the future while managing the present. Not surprisingly, these individuals are snowed over by the competing demands on their schedule. This is especially true for managers asked to oversee change initiatives that rewrite business rules while maintaining the status quo in their full-time job. Fulfilling both roles means neither one receives the appropriate amount of attention, and things fall through the cracks.

Smart leadership teams recognize this misalignment between resources and roles and address it by separating the two. Managers (or process owners) are allowed to focus on the daily grind of their area, including the completion of small, localized improvement efforts.

However, larger efforts are pushed to initiative teams. When managers do get engaged on the big projects, they participate as subject-matter experts. This model accelerates innovation activity—bringing together the resources, knowledge, energy, and focus to execute game-changing initiatives.

Once approved for launch, every initiative needs someone to get the work done. Building the team is the responsibility of an initiative's sponsor. But who is the sponsor? In a process-focused enterprise, I recommend that the owner of the megaprocess most affected by the initiative should serve as the sponsor. In contemporary organizations, one or more seasoned functional leaders usually operate as sponsors. However, my experience is that it is a mistake to have more than one sponsor. Who is accountable for the initiative when there are multiple sponsors? Multiple sponsors equates to diffused responsibility and inconsistent ownership. Additionally, having multiple sponsors fosters an expectation that the initiative team needs their approval throughout the initiative's duration. From my perspective, the sponsor role is not intended to direct the team toward a specific solution. Instead, sponsors should help to organize the initiative and provide executive support but not engage directly in developing the solution. Using this approach, the team has the freedom to focus on building the right solution without the baggage of anyone's preconceived beliefs. If the sponsor's role is to support and not drive the team, the initiative team has increased flexibility to dream up the ideal solution.

Arguably the most critical role to launching an initiative on the right foot is to land the right leader or initiative owner. Leading a "design" initiative requires someone with the flexibility to don many hats: leader, investigator, problem solver, planner, and motivator. Such a person needs to be fact based and objective—yet also intuitive and open minded. As the size and complexity of the initiative increase, so does the need for an experienced and grounded leader. There is no easy way to land the ideal initiative owner—the best candidates surface during a diligent and thorough search. Even then, an individual with the desired credentials may not be available. The alternative is to engage consultants because they often bring additional capabilities, including advanced project leadership skills, third-party objectivity, and experience on related initiatives.

SPONSOR AND INITIATIVE-OWNER RESPONSIBILITIES

With the initiative owner chosen, he or she works with the sponsor to plan and launch the initiative. Table 9.1 shows a sample breakdown of responsibilities between the sponsor and initiative-owner roles. Many of the responsibilities overlap and could be performed by either the sponsor or the initiative owner. As a simple way of identifying the division between the roles, the sponsor builds the team and equips it for battle, and the initiative owner crafts the battle plan and trains the troops.

To align their actions with the initiative's intended goals, the sponsor and owner need to examine all the initiative background materials for completeness. After all, this is the primary direction

TABLE 9.1   Sponsor and Initiative-Owner Roles

Sponsor Responsibilities	Initiative-Owner Responsibilities
Select initiative owner Assist in staffing initiative-team members Communicate initiative to senior leadership and stakeholders across the enterprise Support initiative owner in getting initiative launched	Build initiative approach Staff initiative-team members Reserve space for initiative team Reinforce communications Prepare and train team members Represent initiative to the greater organization and external stakeholders

provided to the initiative team. When gaps or inconsistencies exist in the information, the sponsor and initiative owner should investigate and address the deficiencies before sharing with the initiative team. Ideally, initiative documentation includes at least most of the following components:

- Stated objectives and goals
- Customer perspective (ideally desired process outputs)
- Scope (framed as processes adjusted)
- Strawman of the desired end state
- Anticipated benefits and requisite investments (the business case)
- Assumptions on which the business case and solution are predicated
- High-level initiative timeline
- Business partners and other stakeholders of the solution
- Contingencies and dependencies with other initiatives and events
- Resource requirements to execute the initiative (capital, people, resources, etc.)

When studying any estimates, remember that the initial numbers are only that. Although the business case is necessary to make an informed decision as to if and when an initiative moves forward, the numbers are based on only a cursory design of the eventual solution. The initial business case will with a high likelihood differ from the eventual results—especially with long-term, large-scale initiatives. Use the estimates solely to understand and prioritize the initiative—and then toss them. There is no value in tracking progress to estimates completed before the initiative team designs a detailed end solution. What does it accomplish? Is there a business benefit to validating the accuracy of initial estimates? A greater risk deserving of attention is that the team builds to the initial business case and foregoes evaluations of alternatives. Allow the team the freedom to build the right solution—not constrain it by early conjectures.

Team Member Selection

It is now time to staff the team. The key to a high-performing team is to have a mix of individuals with the knowledge of the subject area and the project skills to get the work done. Finding such a team invariably requires some fishing. As a general rule when identifying potential team members, err on the side of content knowledge. It follows that the areas in scope point to the most suitable candidates. Somewhat counterintuitively, process owners are not good initiative-team members. Their closeness to the subject material brings their biases to the forefront, and their influence is overwhelming. When highly engaged in designing an end state, they invariably stagnate innovation. This is not because they come to the table with preconceived notions (although they may), but mostly because of their emotional attachment to the current process—and this goes for project performers as well. Working with a process on a daily basis breeds an acceptance of its flaws. Embedded work habits are comfortable. Innovation brings change, and most individuals possess a subconscious fear of the unknown. Significant process adjustments may require vastly different skills. A previously superior performer may be an average performer or even need reassignment in a future environment. Although the current process performers provide invaluable information on the current state of the process, an inability to sever their allegiance to the current process makes them a poor fit for brainstorming improvements. As information sources, their input is invaluable, but leave them as a customer of the initiative—not a participant. As a substitute, look for team members from supporting or tangential processes to the scoped areas. Not only are these individuals subject-matter experts, but they also bring a firsthand understanding of how the process's performance affects other areas.

Only a few enterprises maintain a pool of cross-functional resources that are readily available for project work. On rare occasions, you will find an enterprise that values innovation to the extent that it requires employees to take a sabbatical from their full-time responsibilities and participate in an enterprise initiative. But such environments are rare. In most enterprises, there is a mismatch between need for initiative team members and available resources. Compromises are necessary. Out of the gate, aim high. Build a list of ideal candidates, and gradually whittle the list down.

With a slate of candidates in hand, vet each individual to assess his or her interest (passion trumps experience) and what he or she brings to the table. For the moment, ignore availability. The best resources are always stretched beyond their capacity. Some sponsors even tend to make a habit of seeking unavailable resources. At the end of the day, enterprises that value innovation will find a way to free up the best resources. And when resources struggle with the decision to sign up, have the pitch prepared about how the experience will pay dividends for their career.

Benefits to Serving on Initiative Teams (Benefits Increase with Level of Commitment)

- Exposure to other functions/processes
- Learning innovation tools and methodologies
- Opportunity to network with individuals across the enterprise
- On-the-job training to learn new tactics and techniques
- Ability to step back from a current role and identify other areas of interest
- Exposure to subject-matter experts and enterprise leaders

Serving on an initiative team provides a form of leadership training that is unparalleled in the normal course of business operations. When I worked on a major initiative at a Fortune 100 consumer products company, an initiative team included individuals from divisions and functions spanning the enterprise. The aim of the initiative was to build a platform for new-product launches. This required team members to leverage their diverse backgrounds and experiences to craft a process to evaluate new-product ideas and to take the chosen products from idea to delivery. Several years later, I returned to the company to find the original team members were almost all in senior leadership positions. The leadership team may not have planned it that way, but it selected solid performers and armed them with the knowledge and experience to become future leaders.

When forming a team, commit to every team member that he or she will be employed in an equal or greater role after the initiative concludes. It seems stupid, but it is not uncommon in corporate America to find an individual staffed on an initiative, only to see that individual leave the company as the initiative concludes because there is no permanent role available for him or her. Not only is this wasting the valuable experience gained during the initiative, but it also loudly communicates that project work is not valued by the leadership team. When the time comes to staff future initiatives, volunteers will be sparse. This is exactly the opposite of what should be communicated. Initiatives drive innovation—they are the power source for tomorrow's performance. Treat initiative work with the respect it deserves.

In addition to selecting the primary team members, additional stakeholders and periphery team members should be identified and their commitments solidified up front (Figure 9.1). This supporting team includes business partners (process owners), subject-matter experts, technology-team members, supporting-process

FIGURE 9.1 Initiative-team composition.

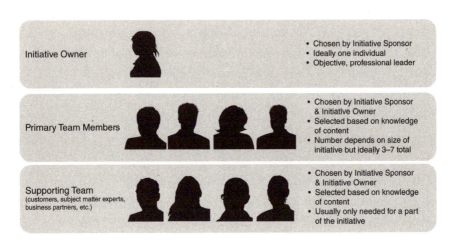

representatives, external partners and suppliers, or just about anyone with something to contribute. By involving them early, they enter the game with the same foundational perspective as the primary team members.

Prior to the official kickoff of the initiative, the selected team members (assuming a full-time project) need to transition all their prior responsibilities to backups. Likewise for part-time initiatives, sufficient responsibilities should be removed from the team members' workload. This is a frequently ignored but extremely important rule. If full-time team members are not removed from their current roles, then they are not dedicated. Their mind and time will be pulled elsewhere (and often directed by their current supervisor who controls their compensation and advancement). The sponsor needs to unequivocally make it clear that team members are expected to be fully devoted. A great way to hammer this home is to make an entirely new department/cost center for the initiative and move the team members into this new cost center—thereby formally severing the link to their prior positions.

PROJECT WORKSPACE

In recent years, the concept of colocating team members has become increasingly popular (especially for full-time initiatives). Colocation (as defined in the Agile methodology) encourages communication, collaboration, and greater efficiency in solution development. Even for part-time initiatives, a team room exclusively for use of the initiative provides significant benefits, including the following:

- It creates a single location to host team activities and eliminate the challenges of constantly seeking available locations.
- It facilitates the collection, storage, and access to information for team use.
- It fosters collaboration and knowledge/information sharing.
- It provides space for confidential discussions (e.g., eliminating roles, challenging sacred cows, and eliminating the political influence of outsiders) where free thought and brainstorming are encouraged.
- It acts as a venue for sharing progress with other stakeholders.

Considerations for a project workspace transcend colocation. How the workspace is organized and used increases the effectiveness of the team. The space and features of the workspace are guided by the initiative's objectives and scope. For instance, a software-development initiative would require workstations to increase interaction between business and information technology (IT) resources.

A key benefit of a dedicated workspace is that it serves as a single place to collect information to be shared. Most project methodologies today burden teams with an excessive amount of documentation. This practice is the unfortunate result of the belief that teams need oversight and control—often to the point of paralyzing progress.

Project documentation is maintained on hard drives, distributed via e-mail, and printed out for review at regular intervals. Some stakeholders will read it; some won't. Some will be engaged in the details and ask meaningful questions; others will grandstand and pontificate. A superior approach for sharing an initiative's status is to post the current versions of designs, work plans, assumptions, business cases, issues lists, and so forth on the actual walls of the workspace. In this way, the team and invited guests can review the information in its entirety without having to search e-mails or shared folders. As the information changes and designs evolve, the wall can be updated. The information is readily available and accommodates impromptu brainstorming and discussion. The savings in time and materials using this approach are immense.

A side benefit of this approach is the ability for teams to set aside time for stakeholders to review the progress of the initiative simply by visiting the room and reviewing the walls. Instead of worrying about formats and customizing presentations for different audiences, the team can provide show-and-tells for stakeholders and partners. While working with a retailer on a project, I developed a tactic that gave further power to the wall approach. As visitors arrived to view the project's progress, the team provided them a pad of yellow sticky notes. The visitors were instructed to write down questions, make suggestions, or provide information. In this manner, their input was immediately captured and placed directly next to the relevant subject matter. Still, even this nonintrusive suggestion approach invites political grandstanding. There are several ways to mitigate influences external to the team. One idea is to set specific hours when individuals can visit the team room and only allow them to use yellow sticky notes to communicate with the team. Also, request that they not include their name or title on their notes. Another idea is to have guided "gallery walks" provided by a team member (other than the sponsor or initiative owner). Remember, these visitations are intended to replace status updates. Continuing to distribute standardized progress reports or conduct status update meetings is duplicative and unnecessary.

Preparation for Initiative Launch

The beginning of an initiative is the most challenging period for the initiative owner. As the day-to-day manager, the initiative owner ensures that the initiative team is fully supported and empowered. The first few days set the tone for the initiative. Getting things off on the right foot requires a good amount of prework, including building an approach for the initiative, assigning initial roles to team members, coordinating training on requisite skills, and facilitating the kickoff of the initiative. The time to prepare for the initiative launch varies, but the owner is always buried with this responsibility for several weeks.

As team members join the initiative, they immediately turn to the initiative owner for guidance. Although the objectives, scope, timing, and general expectations are shared during the team member selection process, team members often struggle with the ambiguity that is the essence of initiative work. To provide the appropriate amount of guidance and build momentum, the initiative owner should overorganize and overcommunicate in the early days. While the team is still being built, planning should be well underway for the kickoff meeting, any training sessions, and the initial weeks of the project. If outside resources are engaged for training or tours and informational sessions are required, the arrangements should be locked down and confirmed.

As the kickoff approaches, a communication plan is needed to drive alignment across stakeholders, business partners, and the remainder of the enterprise. The communication plan includes multiple layers of communication—each customized for its intended audience. Distribute a general communication to a broad audience to communicate the initiative launch, the primary goals of the initiative, and whom to contact with ideas or suggestions. To substantiate this message, a slightly deeper communication with frequently asked questions (FAQs) should be distributed to senior leaders and megaprocess owners to answer employee questions that might arise. The most detailed communication is reserved for the initiative's business partners and stakeholders. This communication includes a high-level timeline and the expected commitment from business partners and other stakeholders. All communications should be ready prior to the kickoff to ensure consistency in the message from all team members.

INITIATIVE KICKOFF

The initiative kickoff is the first time every team member is present. The sponsor and the initiative owner jointly own this meeting, although they must approach the meeting from different angles. The sponsor communicates the background and objectives—giving context to the quest. The initiative owner speaks to the logistics and the approach to be employed to achieve the initiative's goals. Generally, the kickoff is exclusively for the initiative team. The primary goal is to develop a shared foundational understanding of the initiative and to lay the groundwork for its completion. Topics covered during the kick-off include

- *Initiative objectives.* The explicit and implicit needs to be met by the initiative as designated by the leadership team/ committee.
- *Scope.* The processes expected to be affected by the initiative, although the scope should not limit the team in developing the correct solution.

- *Initiative approach.* Detailing the manner in which the team will work to execute the initiative. This is covered in the latter half of this chapter.
- *High-level work plan.* Bringing the approach to life by identifying the specific activities and their anticipated duration. The work plan also identifies key milestones over the course of the project.
- *Roles and responsibilities.* Identifying the roles each team member will fill. The specific role for each team member should be shared prior to this meeting so that there are no surprises. This agenda item is to communicate this information to the full group.
- *Communications and completed activities.* A review of the communications sent previously or to be delivered that outline the team objectives—as well as any completed activities that have been done in support of the initiative (e.g., FAQs, general communications, team member scripts, and scheduled meetings or tours).
- *Work rules.* Governing rules for the team, including how decisions will be made, how issues will be resolved, how the team will work together, and general guidelines and procedures for the team.
- *Building the wall.* Placing all known initiative information and work in process on the walls of the team workspace. This activity invariably leads to team chatter and discussion on the initiative and acts as a fire starter to build team momentum. Remember, the wall is workspace, and the documents, diagrams, and charts are meant to be annotated and updated with progression of the initiative. Key items to display on the wall include
  - ▲ High-level work plan
  - ▲ Business case evaluation/assumptions with regular review

- ▲ Milestones and checkpoints (e.g., financial reviews, management reviews, and stakeholder reviews)
- ▲ Project assumptions (eventually expands to solution assumptions)
- ▲ Stakeholder and key contact lists
- ▲ Ongoing resource adjustments/increases
- ▲ Issue management
- ▲ Risk management
- ▲ Team rules—especially to make decisions and handle conflict
- ▲ In-process work (e.g., brainstorms, items to remember, and issues to resolve)

Depending on the team's composition, there may be skill or knowledge gaps to be filled. In the first week or two, squeezing a training session or two into the schedule is usually relatively easy. Training at the onset emphasizes skills of importance to the collective team. Examples of training classes conducted on real projects are provided in the following list. Most of the classes focus on skill development, but functional area training may be included as well. Functional area training focuses on areas such as the supply chain, retail operations, European business practices, and others. Down the road, follow-up sessions can be conducted as needed. It goes without saying (but frequently needs to be said) that training should focus on critically important skills that will assist the team in completing the initiative. Training for the sake of training is a waste. Examples of training classes include the following:

- Basic process training, including flowcharting, process analysis, informational interviewing, process design, testing, piloting, and financial modeling
- Specialized process training, including Lean, Six Sigma, and process transformation

- Change-management training
- Customer analytics and customer feedback channels
- Technology including specific software packages
- Knowledge training on specialized functions inside the enterprise (e.g., supply chain, operations, sales, customer service, etc.)

At this point, the initiative is proceeding forward and beginning to gain momentum. The team is engaged and enthusiastic, and the initiative owner transitions from leading the team to becoming a supportive parent. More than anything, the initiative owner sets the tone for the team—keeping morale high and ideas fresh. He or she needs to be a supporter and facilitator, never a naysayer or dictator. The importance of this role cannot be understated in moving the team forward in the accomplishment of the initiative's goals.

REVIEW THE GOALS OF THE INITIATIVE

The continued existence of any process depends on it fulfilling a business purpose. Likewise, an initiative is created to align a process or group of processes to meet that need. The need—whatever it is—is the logical starting point for any initiative. As one of the immediate tasks after the team is assembled, I recommend that team members collectively review the documented rationale behind the initiative. The intent of this exercise is to begin exploring the answers to a handful of fundamental questions (and optimally to do so in process terms).

- What is the desired end result?
- What is the relationship between the process(es) in scope and the overall enterprise?
- How is the enterprise's strategy embodied in the process(es)?
- If process requirements are available, what are they?

- Who are the customers of the process?
- Who are the other stakeholders?
- What are the goals for cost? Safety? Quality? Throughput?
- Are there other requirements of the process?

While debating these questions, the team should construct a statement (or series of statements) that communicates in their own words the initiative's intent. The aim is to quickly get the team aligned and singing the same tune. When the team reaches agreement, post the statement prominently in the workspace for future reference.

An Initial Future-State Design

Once the team has a shared perspective of its target, team members I recommend immediately get the team engaged in thinking about what the end solution might look like. For the moment, ignore any known constraints (because are they really constraints?). Just get ideas down on paper! Even better, create multiple alternatives to explore. Although rarely will several solutions for the same problem be fully analyzed, there are minimal risks to an abundance of ideas, and the potential benefits are huge. Why not have multiple solutions to vet?

On occasion, I am asked the rationale for encouraging teams to think about a final solution prior to completing any due diligence on the current processes and environment. There are two reasons. First, nothing engages a team like asking team members to draft a solution on a whiteboard. This simple exercise immerses them in the details of the problem and forces them to think. It is the ideal ice breaker—jump-starting idea generation while simultaneously forcing team members to consider unbounded options. Second, any design built at this early point is unfettered by the biases of leaders, process performers, or business partners. Before the "That will never work" attitude is injected (and it will be), team members can linger in a blissful vacuum for a bit and let their creative juices bubble. Encourage them to consider everything and anything at this point, including expanding (or limiting) the initiative's scope. Entertain those crazy ideas that just might turn into something magical.

While everyone has an opinion on what works and what does not, very few individuals can jump onto an initiative and immediately start designing a future state. Often teams will spin their wheels and delay moving forward because they are new to the ambiguous realm of process design. Symptoms include a continual review of initiative documentation, reclarifying existing initiative details, or failing to get any ideas on the board because they are just not perfect. Teams often struggle when there is not a conveniently paved road for them to follow. Under these circumstances, I recommend the team take a final moment to review the initiative details (i.e., output requirements, voice of the customer, and any other relevant data), and then put them away in the file cabinet for the moment. Then just start brainstorming. Go around the room and require everyone to put an idea on the board. Think about what the end state might look like. How can the customer's experience be changed? Play with the process. Put crucial parts of the process on yellow sticky notes, and adjust their order to discover new ways of performing the work. Challenge every step to see if it could be done differently. Draw arrows to indicate the order of steps. Seek originality. Revisit critical points in the process. Just get started!

If the team is still stuck, explore the process models (found in the Appendix) and investigate their applicability to the situation. Process models are proven ways to organize work to achieve different outcomes. For example, if a process requires a number of specialized skill sets, one alternative model is the *caseworker model*. Frequently employed in loan processing, this model uses a single individual (a caseworker) to manage the flow of work between specialized performers. The intent of using process models is to consider different ways the work can be performed to arrive at improved outcomes.

Beyond the process models, another helpful tool is to consider the initiative from various vantage points, including the customer, business partners, or the performers. Questions such as the following are useful to get team members to consider different perspectives:

- *Review the process from a customer's perspective.* Is the end result challenging for the consumer? Frustrating? Or will it delight the customer? What more can we do for the customer?
- *Analyze the available customer data.* What is the data telling you the customer appreciates? What don't they like? How can we delight the customer?
- *Build the solution to the problem.* How can the process design be simplified? Can anything be eliminated? Performed elsewhere? Don't overengineer and solve world hunger. Keep it simple. Is every step really needed? Don't let perfection be the enemy of good? Beat the competition— don't build utopia.
- *Start with the initial state.* What would improve the initial state at this point? A greater focus on efficiency? Less cost? Higher quality? Is there a competitive advantage to be gained?
- *Think of additional information or answers that might change how the process is constructed.* Make a list of questions to investigate. Is every output of the process needed? Are the inputs flexible? Are there a lot of exceptions to the process? Are there unintended consequences of the process?

While thinking through the initial-state solution, set up a place to collect and eventually answer questions. Often items are uncovered that require additional effort to resolve or build into the solution. Do not allow these questions to get buried in e-mail chains or forgotten in meeting notes. But also make sure that they do not hold up progress either. Write the questions in a convenient location (i.e., a parking lot in the project room), and set a time to return to them in the future. In many instances, they resolve themselves with the passing of time.

Also remember that in some instances it is not appropriate to document a process. As stated previously, this occurs when there is variability in the inputs, outputs, or the process itself—and this variability or uniqueness is valued. Under these conditions, create standards or procedures to frame the work effort. The aim of work organization remains the same—complete the work to improve the customer experience and do so with a reasonable return to the enterprise's stakeholders.

It is worth repeating that during this brainstorming phase, the more ideas the merrier. The initial state is but a stake in the ground. The true benefit of this activity is to engage team members and have them think through the challenges and opportunities inherent in the initiative. Although a design or designs exist, team members always discover that there are gaps in their understanding of the situation. They need more information to design the optimal solution. This brings us to the next step—formulating a plan of attack to plug the gaps in their knowledge.

CURRENT-STATE ANALYSIS

The current state represents what happens in the process today. Understanding an existing process requires a thorough analysis of its current operation, including the process itself, the customer, the inputs, the outputs, and any control mechanisms. Once known, the current-state process is the baseline to compare against any new designs and measure improvements. It identifies the inputs and outputs, provides a frame of reference for metrics, and allows insight into the customer experience. The act of pulling this information together helps to identify what is missing in the initial state. Is a deliverable missing? Are there constraints to be incorporated? Is what was devised in the initial state even feasible?

Through an understanding of the current state, team members gain the background understanding to ensure that they build a process capable of accomplishing the initiative's goals.

CURRENT-STATE DOCUMENTATION

Over the past few years, a number of "recommended" formats and notations have been developed to document processes. *Business-process modeling notation* (BPMN) is one example. In general, most documentation methods are fairly similar. Although I applaud the goal of standardization, process documentation does not need to be translated into any specific format. As long as it is understandable by a wide audience, the format is sufficient. What is vitally important is capturing a complete view of the activities that together comprise the end-to-end process. When determining the start and end of any process, be expansive and aggressive. Try to include every step and activity that influences the value derived by the process. Whereas an overly large scope may force the breakdown of work into manageable pieces, having the ability to smooth the flow of work from end to end expands the opportunities and the potential outcomes.

For the most part, documentation is discovery—sifting through the innards of the process to see how the various pieces fit together. Documenting is akin to an archaeological dig. You excavate to uncover the general shape, hand dig the next level, and then sift to get the details.

Excavation includes conducting interviews with process performers, business partners, and other stakeholders. Although invaluable information can be captured quickly through interviews, it does come filtered by an individual's vantage point. It may be tainted by a personal perspective and not provide a true window to reality.

Continuing with our archaeological metaphor, digging by hand is equivalent to observing the process in action. It expands on the foundational view gained through informational interviews and builds depth to our understanding of the process. Perhaps the greatest advantage of observation is the ability to differentiate the theoretical from the actual, and almost as important is the ability to see exceptions and how they are handled. Exceptions are quite frequently unknown or glossed over during interviews, but they need to be noted when creating or adjusting a process.

Finally, sifting through the soil on a dig equates to the hands-on execution of a process. At this level of discovery, the investigator puts his or her "hands in the dirt" and actually performs the work like any other employee. Executing the process firsthand deepens the investigator's understanding and allows him or her to pick up details not previously discernible.

Using these techniques, the details of the process emerge, giving the investigator the ability to see how the puzzle fits together. A well-documented (and understood) current state includes the following elements:

- *Outputs*. What are the products or services created? What are the attributes of the outputs? What is important to the customer (i.e., process requirements)?
- *Inputs*. What are the knowledge or raw materials required to run the process? What is the flexibility in changing or replacing them?

- *Constraints.* What are the constraints on the process (from business partners, industry conventions, or legal/regulatory requirements)? Are they flexible? Can they be reduced or eliminated (these are important to document as assumptions of the initiative)?
- *Process steps.* What are the activities in the process? In what order are they performed (e.g., sequentially, in parallel, dependencies, no connectivity, alternative paths)? Has the order changed over time? Why?
- *Performers.* Identify the individual who executes each step in the process. How are the handoffs managed?
- *Decision points.* Does the process accommodate variations? What are the decision points leading to alternative paths? Are there exceptions? How are they handled?
- *Process metrics.* What metrics are used to track performance or control the processes execution (i.e., throughput, inventory, time in stage, etc.)?

Again, documenting the current state is a discovery process—when thoroughly done, it uncovers a wealth of valuable information. Unfortunately, there are no shortcuts to capturing this level of detail. The investigator must get their hands dirty.

CUSTOMER, INPUT, AND OUTPUT ANALYSIS

Customers are the reason for the existence of the processes in any enterprise. What customers want and what they dislike determine whether they will purchase an output. The intent of an initiative is to either improve an output to make it more desirable to customers or to expand the existing process capabilities to deliver even better outputs in the future. By retracing the steps of the current state, you can determine the full set of outputs and customers of a process. This is accomplished by asking a simple question at each step: "What is delivered by this step, and to whom it is delivered?" Using this approach, you can identify outputs that are created throughout the process's execution as well as the final delivery to the end consumer. Many processes make outputs for other processes or functions. When an output is ignored, internal customers and business partners may be negatively affected by a process-improvement effort.

In the same manner, the process needs to be examined step by step to capture the inputs. At each step, ask the following questions: "What is needed to complete this step? Who delivers this input? Note the specific input and its quantity or quality. Are there any alternatives? For example, when the primary supplier is not available, can inputs be sourced from another provider to keep things moving?

At this point, the required due diligence is complete. The initiative's goals are known (we know where we are going), the current state has been explored thoroughly (we know where we came from), and an initial state is complete (we have a blueprint to build from). Now the initiative team is in a place where many falter—tripped up by a widely held belief that is largely false. Contemporary business theory suggests that process innovation—really, any improvement—needs to be metric based. In other words, improvements need to be measured—or so the thinking goes. Given the widespread adoption of metric-based improvement goals, a discussion on process innovation would be incomplete without addressing this practice.

METRICS AND BENCHMARKS

If you work in a corporation long enough, you invariably will hear the adage, "What is measured gets done." I will argue that a related statement is far more truthful: "A metric tracked to reward individuals is almost always achieved." A metric is, in fact, an excellent indicator of a performer's focus. This does not necessarily mean that the process is aligned with the strategic aims of the enterprise, nor is it creating products/services that are desirable to the consumer. It simply means that some metric is being achieved. And here is where measurements can derail progress. The metric may not be the correct metric, and its achievement may not even be desirable to customers. And, of course, there is the situation where the metric is gamed by the performer and achieved, although not in the manner anyone intended. For example, to cut costs, a process owner may buy an inferior quality of raw materials. The cost-reduction goal is achieved, but the product is now inferior to those produced previously. And there may well be additional costs because the inferior inputs require additional rework to manufacture the products. Additionally, the lesser quality will likely result in a higher incidence of customer-service issues. Many business leaders are finally beginning to understand that metrics may not be the best way to frame improvements. If you really want to change the way work is completed, focus on the processes, not the metrics. To many, this means that the process needs to be benchmarked—a related business myth.

When analyzing processes, the topic of benchmarking always surfaces, and the discussion moves quickly on how to obtain benchmarks and map performance against the averages. A better question is whether benchmarks should even be used when analyzing any process. Answering this question appropriately requires differentiating between the source of the benchmarks—externally provided and focused on a specific industry or those benchmarks generated inside the enterprise. Externally provided benchmarks are captured and distributed by research companies or institutions. The research firms gather the information at its source and compile it for a fee. In general, obtaining and using external benchmarks to identify improvement opportunities is a step in the wrong direction for several reasons.

Benchmarks are an average of a number of companies. Do you really want to map yourself to an average? If you are the industry leader, why benchmark the competition? If your goal is to outperform the competition, why the interest in the average?

In addition, companies operate in different environments and possess different strategic and operational structures. The environment (e.g., cost of labor or cultural differences) will distort metrics—and therefore benchmarks. Using a benchmark to contrast performance is like conducting the same science experiment and heating the substance in one experiment and freezing it in another. Could you expect the results to be the same? Environmental conditions matter.

The real underlying reason for benchmarking is to identify opportunities for improvement. Even if there is perfect alignment in the situational factors (i.e., environment, strategy, geography, culture, operation model, etc.) between a benchmark and the enterprise in question, what do you do after the variance is identified? The reason for the difference is not conveniently listed with the benchmark—so what was the point of benchmarking? Are you really better off than if you skipped the cost and time associated with a benchmarking study?

Internal benchmarks are a different animal altogether. Internally generated benchmarks come from internal tracking systems. The information never leaves the walls of the enterprise. Many companies with multiple manufacturing facilities, distribution centers, stores, or other similar-use locations may benefit from benchmarking. The difference here is that discrepancies can be identified and investigated as to their root cause. If one facility is operating at a lower cost, the reason can be explored and potentially implemented at other locations. Even when using internal benchmarks, however, care should be exercised to ensure that the benchmarks are appropriate for comparison.

As a last word on metrics and benchmarks, my recommendation is to be extremely careful setting metric-based goals because they will become the focus of the process owner and his or her team. Make sure that the metric is valued by the customer and is measured across the full end-to-end process. Additionally, only use benchmarks when they are internally generated and reflect the same environmental conditions as exist in the subject process.

ITERATE AND REFINE THE INITIAL STATE

The initial state is but a conceptual view of an end solution. With the information gathered during analysis of the process, the initial whiteboard designs usually need some updating. But getting the process right requires more than layering this information into the designs. Process creation and improvement require trial and error. There is simply no substitute for experimenting and seeing the results. The intent of the "Iterate and refine" phase is just that—to incorporate available information and play around with the process until a reasonably complete solution surfaces.

The first rule is simple: get started now. Initiative teams waste days in search of an ideal starting point—a fruitless endeavor to make the process clean. Forget it. Get dirty! Put the pencil to paper, and get rolling. At one of my employers, an often-expressed quote in the process laboratory was, "You can't get there from here, but you can get here from there." Although somewhat confusing, the point of this quotation is that you may not know enough to design the perfect solution now, but as you test the boundaries, new options become visible.

On occasion, the initial-state designs created previously are discarded as new information becomes available. When this occurs, some process experts suggest initial states to be worthless activities. But even if the initial state ends up being complete trash, the exercise engaged the team and revved the creative engine. In the absence of an initial state, teams tend to limit their ideas to perceived boundaries (whether real or imagined) and never take the time to dream up possibilities before the hard data arrive. Use the initial state to be bold and expansive. Spend time hunting for that game-changing innovation. On occasion, you strike out, but there are times when the ball flies 400 feet and you win the game. Swing for the fences!

Even when the initial state generates a dozen options for further inspection, continue to encourage the team to pump out new ideas. The initial state provides a foundational perspective on where to go, but new options and opportunities surface during the analysis phase. Let them flow. This is a great time to get the full team together and scrawl new designs on whiteboards. Ask the basic questions: Who? What? When? Where? Why? How? Break down the walls. What if industry conventions and rules were no longer valid? Can a new business model replace the existing one? Can the boundaries be pushed out to provide greater customer options? Be bold. You can build a revolutionary mousetrap!

During this period of free-form ideation, process constraints inevitably will surface. Some of them are legit and require compliance, such as government regulations. Others, including industry conventions, were instituted by previously dominant competitors and are outdated and ripe for eradication. Identify them. Categorize them based on their ability to be overcome. Some constraints, including governmental regulations, universal standards (e.g., the long-term use of the English system of measurement), continued practices (e.g., use of cheap labor), and locations of facilities, defy easy adjustment. Once the constraints are named, shuffle through them as a team, and gauge whether they are worthy of being attacked and eliminated. Sometimes, they are easy to discard. At other times, the juice is not worth the squeeze—especially with industry-accepted constraints. Remember, in competitive markets, the goal is to beat the competition, not provide the ideal solution. The competition is encumbered by many of the same constraints. Other innovation opportunities may prove easier to bring to fruition and with a greater payoff.

When the process design(s) delivers the intended results, switch gears and focus on adjusting the process (or create new versions) by manipulating what can change. A convenient place to start is with the deliverables—a.k.a. the process requirements. Reexamine the customer's preferences, and investigate ways the outputs might be expanded or adjusted to provide greater value to the customer. Cull the customer research and feedback from the front lines for latent customer needs or desires. The point of this exercise is to test the flexibility of the process. Often the best approach is just to brainstorm and list the options on the project wall. Can a service be tied to the product? Could the product be enhanced for specialized uses? Can the product be made at a cost that is low enough to appeal to a wider audience?

Continue this same exercise with the inputs. In most instances, inputs do not provide the same amount of flexibility, but they are still worthy of exploration. Adjusting inputs may alter the quality of the product or provide expanded functionality. For example, the development of a new chip processor drove innovation in the iPad. Are there alternatives that could be used in lieu of the current inputs? Are there other potential suppliers for the inputs? What are the substitutes? Each of these questions gets to the root of how deliverables might be adjusted. Examine the effects and determine whether they are desirable to the customer and reasonable for implementation. Again, listing them on the project wall ensures that the options stay top of mind and continue to be considered as the process evolves.

Finally, examine the process itself. Use the process models to drive brainstorming.

- Can the process be changed to expand the customer relationship?
- Is it possible to customize the product/service for a specific customer group?
- Can products be bundled for the customer?
- Can the product/service be adjusted for a new market?

The process itself drives the outputs in the same way the inputs do. The difference is in the magnitude of the changes. Rearranging the steps in a process, adding new elements to the process, or simplifying the process may create change that echoes throughout an industry. You need only to think of the different processes used by companies such as Dell, Amazon, or Apple to see how industries were revolutionized when processes were adjusted to solve customer pain points or to deliver a better customer experience.

As mentioned previously, innovation is an iterative activity that requires injections of creativity brought on by new perspectives and outside participants, including other employees, customers, suppliers, business partners, and so on. Sometimes the ideas just need time to percolate. Do not rush the process, but also recognize that it must end. With a complete exploration of the outputs, inputs, and processes, several solid solutions usually result. The question is when to stop. When is enough enough? The answer is microwave popcorn. When the popping slows to a trickle and ideas are no longer jumping out, it is time to put the existing designs to the test. It is time to put them in the laboratory to see how they perform.

LABORATORY TESTING

In 1990, *Parade* magazine included a thought-provoking exercise that drew the ire of a few professional mathematicians. It was presented in the form of a game show where an audience member is shown three doors. Behind one of the doors is a brand-new car, but behind the other two doors are angry goats. The participant is allowed to select one of the doors. She selects door number one, but it is not opened. The host then opens door number two to reveal a goat. The host then asks the participant if she would like to change her selection from door number one to door number three. What is the correct response?

The most common response is that it makes no difference. This answer is also wrong. The true answer is that switching is the best bet. It increases the chances in favor of the participant from one-third to two-thirds. The human mind struggles with the logic because the results seem counterintuitive. Only when the results are diagrammed do any respondents begin to see the logic. Even then, many still struggle and actually must simulate the game and track the outcomes. I have used this example in dozens of lectures and have yet to find someone who answered it correctly on the first shot. The point is this: as smart as we all think we are, life often befuddles us when results do not follow our expectations.

For this reason, we cannot assume that the designs created by the initiative team will perform in the real world. The best way to determine what works and what doesn't is to test the designs in as real an environment as possible. This is the process laboratory. Most (if not seemingly all) initiative teams ignore this vital step—much to their later chagrin. Every proposed solution should be tested within reason based on the size and scope of the problem/solution. Do not skip this step!

The intent of the laboratory is to simulate an environment as akin to the real-world environment as possible. Although a laboratory will never perfectly predict actual performance, it may very well provide results that lead to adjustments or a clear repudiation of a solution design. To maximize its predictive capabilities, the laboratory should replicate actual environmental conditions and utilize anticipated use cases. When possible, the performer of the process should be an actual performer. The products and services should be produced as they would during the normal course of business. If possible, actual customers should be used to establish the true spirit of the interactions. And although often overlooked, real systems and tools should be used whenever possible. In lieu of developing the supporting systems, prototypes or simulations are an acceptable substitute.

With everything in place, begin the laboratory experiments. Test actual and predicted business situations from beginning to end. Repeat each use case multiple times. Log the results. Then test the same business cases again, adjusting the known variables. Again, log the results. While testing, incorporate disruptions, disturbances, and interruptions to mimic real-world situations. For example, test how the process works when timelines are compressed or when disruptions (such as a late employee) affect the process. The laboratory is not intended to be a sanitary environment. Laboratories provide the best insight into solutions when they are fully saturated with the stench of reality.

General Guidelines for Conducting a Process Laboratory

- Replicate the actual conditions and environment to make the laboratory tests as realistic as possible. Use the actual process performers and use cases to test the proposed solutions. Introduce variability and disruptions known to exist into the modeled environment.
- Build the structures and prototypes (e.g., timed screenshots to simulate software solutions) that support the process

- Work through each design thoroughly—from start to finish. Repeat it multiple times while simultaneously injecting variability into the iterations.
- Track performance. Identify the process designs that perform the best. No process works in all situations. Be willing to accept failure. Identify the risks, and mitigate them, if possible, that is, if the risk is prevalent enough to warrant a response (e.g., the tidal-wave risk in Kansas is relatively low).
- Capture feedback on the tested designs from multiple perspectives: customers, business partners, suppliers, outsiders, line-level workers, and so on.
- Iterate and adjust. Continually reevaluate. Repeatedly ask what can be improved. Try substituting performers with different skills. Try different environments. Change the tools.
- Document exceptions as they are encountered. Identify ways to handle them, and complete a cost-benefit analysis to see if the juice is worth the squeeze. Should the process accommodate exceptions? Or do they ruin the business case? Let the competition serve unprofitable customer segments. At the end of the day, there is no perfect process—exceptions persist in flawlessly designed processes. Live and let live.
- Bring in fresh eyes. Listen to their feedback. Avoid the proverbial road to Abilene where everyone agrees on the surface, but unvoiced misgivings exist.
- Continue to refine the process until it is good enough. Use the rule of microwave popcorn.

When do you stop testing and begin planning a pilot launch? Here are some guidelines:

- Balance getting the process out there with getting it perfect. Don't let perfection be the enemy of good. In fact, perfection is rarely appropriate as a goal. In competitive markets, all that is needed is an advantage over the competition.
- Quit testing and get a pilot out there when the solution meets the output requirements and generates value over what exists currently.
- Timelines may need to be hastened if the competition is launching a similar product. In such circumstances, *never* release an inferior product/service unless it is proven to be of value to the consumer (the competition may have overbuilt) and it has a significant cost advantage. Be extremely sensitive in these circumstances. Inferior products may stain the enterprise's reputation. Make sure that the value exceeds that of the competition's offering.

Before piloting a solution, revalidate the business case. Although the business case never should be far from the initiative team's mind, this checkpoint ensures that a solution is not piloted that does not make financial sense. If there are outstanding questions or unconfirmed assumptions, make sure that they are captured so that they are addressed when additional information becomes available.

Lastly, before piloting, take one final shot at making the process more efficient. The main goal of process innovation is to deliver the right customer-focused solution, but time spent driving efficiency in the initial solution often pays dividends. Consider using one of the popular process-improvement tools, such as Lean or Six Sigma. Use Six Sigma when the goal is to deliver the highest quality of products/services; use Lean to improve speed to delivery, quality, cost, and safety. Complete this exercise for all the solution designs. It is now time to pilot the solution(s). Although frequently one solution appears to be far superior, it never hurts to pilot alternatives.

PILOT

The intent of the laboratory is to make mistakes before putting the solution in front of a large audience. But no matter the extent to which the laboratory replicates actual circumstances, the pilot is the first time the solution is employed in the real world. The immediate decision is where to pilot the solution. I recommend to host the initial deployment at Main Street, USA. In other words, pick a rather generic part of the market where general learnings can be gained—but with minimal unique characteristics. The pilot customer/location should be representative of the real world but isolated to the extent that it limits the risk of failure. A convenient location for monitoring the solution is also a consideration because pilot designs nearly always require adjustment prior to an expanded deployment.

Preparing for the pilot entails many of the same activities undertaken in the laboratory. Employees are trained on the new process. Tools, systems, marketing materials, and other elements of the solution are acquired/built and set in place. Facilities are reorganized and retooled. Communications are routed to business partners, suppliers, supporting teams, and other affected parties.

With the infrastructure in place, begin the pilot. For the most part, the initiative team just observes. Unless a part of the solution is not functioning as designed, let the process play out—even though it may be painful to watch. Remember that this is the first time these performers have used the new process and system. Perfection rarely visits during the initial iteration. Resist the temptation to tinker until the pilot has completed several cycles. Even then, refrain from adjusting the process for anything other than addressing critical issues in the initial location. Wait to address minor issues until the solution is expanded to additional locations/customers. What is a minor deficiency may look entirely different in a new location. Expanding the pilot creates an opportunity to either adjust the original design or try an alternate. Additionally, it provides data on how the solution fares in different environments.

Monitor and treat both the initial pilot and the expanded pilot as if they were still in a laboratory environment. Track performance, and as time passes, slowly begin making adjustments. Be patient. When the solution is working and the results are meeting the goals, it is ready for deployment.

DEPLOYING THE SOLUTION

With the completion of a successful pilot, it is time to build the organization and supporting structures for a general deployment. Because the pilot was limited in scope, building the long-term components requires additional effort. There are two primary deployment approaches—a *direct cutover to the new process* or a *staggered deployment*. The direct-cutover approach replaces the existing solution via a one-time event. It takes significantly less time and allows for the immediate ramp up of benefits, but it comes with greater risk. A staggered deployment is more cautious. It replaces the existing solution in select locations and gradually expands to other locations over time as long as there are minimal issues.

The approach to deployment should balance the benefits with the risks. If time to market is the preeminent aim and the pilot was successful with minimal adjustments, direct cutover is a viable option. If there is substantial risk owing to variances in the environments where the solution will be deployed, a phased or wave approach is more appropriate. Also remember that the initiative is not operating in isolation. Other initiatives may be deploying solutions to the same area(s). In these instances, working with other teams to collaboratively deploy the solutions may minimize disruptions to the business. All these factors should be considered when building the plan.

Regardless of the approach chosen, a well-executed deployment depends on several critical elements:

- *Initiative deployment office.* A team to plan the deployment, ensure that all pieces are in place prior to going live, and with the authority to make adjustments on the fly.
- *Facilities, resources, and tools in place.* All facilities, inputs, tools, and other components of the solution are built to expected volumes and are scalable and flexible enough to accommodate potential adjustments.
- *Communication.* The solution is fully communicated to all business partners and stakeholders affected by the new solution. Their roles are communicated and understood.
- *Change management.* The performers and other individuals involved in the process are aware of the change and prepared to perform their new or adjusted roles. This includes the creation and execution of a training program to provide them with the skills to function in the new environment.
- *Feedback loops.* When the deployment occurs, the appropriate communication loops are built and in place (i.e., reporting, conference room discussions, etc.). This allows for the identification and escalation of issues when discovered (performance or situational).

TRANSITION PROJECT ROLES TO ONGOING PRODUCTION

With deployment of the solution across all locations, customers, and appropriate product lines, the initiative team is nearing completion of its work. At this point, many of the team members may shift into new roles, and only a few of the original cast may remain.

Although this approach is not metric based, businesses operate to make a profit, and other organizations operate to fulfill a purpose. Invariably metrics come into play and are a part of the monitoring and adjustment of the solution as it moves into steady state in a production environment. As the solution is turned over to the frontline performers, metrics allow for the tracking of performance versus the current state and allow for the identification of opportunities and issues. From my perspective, the accounting system financials and basic metrics (i.e., output, time in process, and market share) are sufficient to report on progress. That said, I have never witnessed a solution deployed that did not come with a ream of reports to monitor and track progress. Again, I argue that most of this paperwork is wasteful. Processes ought to be managed at the process level where the work is completed, not off reports that cannot possibly identify the myriad influences and unique characteristics of the actual environment. In lieu of fighting this battle, I suggest a sunset period for extraneous reports. As the initiative team passes off the management and reporting of the initiative to the full-time team (i.e., the process owners), identify a time when these extra metrics and control processes can be eliminated. If new issues surface, new reporting can be initiated. Even in these instances, ensure that the cost of reporting is exceeded by the benefit gained by tracking. Enterprises spend a tremendous amount of time, energy, and money reporting on everything and anything. The reporting volume far exceeds the time available for employee to absorb and act on the information. In general, this time and effort would be far more productive if it were shifted to customer's requirements and used to fuel the innovation process.

With the initiative complete and the transfer of ownership to the front line, a few remaining activities should be completed to close down the initiative:

- Ensure that the solution is documented in training and other corporate information repositories.

- Hand off the remaining project work documents, including alternative and potential ideas for the process, to the process owners. This is often a wealth of information that, if used, could shortcut future initiatives.
- Update the final business case results.
- Share the results of the initiative and lessons learned with the greater enterprise. Create a case study of what happened to allow for collective knowledge and awareness. Many teams suffer through the same challenges in building and deploying solutions. Sharing experiences helps the enterprise to institutionalize competence in innovation.
- Document/report on the performance of external team members (i.e., consultants, contractors, and others), and provide this information to the appropriate individuals.
- Release any materials, facilities, systems, or resources no longer needed back to the enterprise resource pool.
- Celebrate success!

The last point—Celebrate success!—is more than just a feel-good point. It may very well be the most important bullet point in the preceding list. The less prominent initiatives in contemporary corporate America tend to wither into the background once the initial luster dissipates. Inadequately staffed, underfunded, and with disinterested sponsors, they linger on portfolio management lists but are the walking dead of the initiative world. Celebrating success is about building enterprise momentum to support initiatives and see them through to their successful conclusion. This is not to say that there are not those occasions when the market changes and a portfolio-management function appropriately weeds out the laggards. But initiatives generally deliver value when they are adequately resourced and executed to completion. At the finish line, a celebration recognizes success and encourages others to push forward. Value is created, and the strategic and operational positions of the enterprise are improved.

10

A Process-Focused Enterprise in Action

G rasping the significant differences between a process-focused enterprise and a traditionally operated business requires one to be immersed in the ongoing operations of both models. Because most individuals today work in a traditional environment, what is missing from their education is a peek behind the scenes of a process-focused enterprise and how it handles an initiative from its inception to its execution.

To this end, let us explore a manufacturing company (WidgetCo) that builds the proverbial widget. For most of its history, this market has been relatively stable and profitable. Roughly five years ago, this lucrative market attracted the attention of international competitors—and several entered the market by offering a lower-price widget targeting the mass market. More recently, though, the widget market underwent a downturn. Total market sales fell to roughly $150 million from $250 million, causing several smaller competitors to get pinched and either go bankrupt or get acquired by others. After the dust settled, only one foreign company (ForeignCo) remained, and the domestic market included two large competitors in WidgetCo and BigGuyCo and a handful of specialty manufacturers, including SpecialCo. But there are signs of life. The sporting segment, which previously tallied sales of $15 million annually, grew substantially last year—much to the benefit of WidgetCo and BigGuyCo. BigGuyCo offers a beginner model in the sporting market, but as players' skills grow, they eventually upgrade to WidgetCo's sporting performance line. The current state of the market is shown in Figure 10.1.

Several weeks ago, BigGuyCo grabbed the headlines of industry journals. Choosing to focus on its core business line of outdoor power equipment, BigGuyCo announced that it was shuttering several business lines—including widgets. Because BigGuyCo is WidgetCo's major competitor in the sporting segment, BigGuyCo's exit creates a major opportunity for WidgetCo to nab market share. As soon as WidgetCo's strategic planning team got wind of the news, it began to brainstorm strategic maneuvers to capitalize on

FIGURE 10.1 Market map for widget industry (each widget = roughly $3 million in sales).

Customer Segments		Providers				
		WidgetCo	BigGuyCo	SpecialCo	ForeignCo	Customer Summary
	General	(7 widgets)	(7 widgets)		(13 widgets)	• Uses product in varied manners • Cost focused • Appreciates minor customizations
	Sport	(10 widgets)	(6 widgets)	(1 widget)	(1 widget)	• Not as cost focused • Appreciates performance (lighter, greater strength)
	Leisure/ Customized	(3 widgets)		(5 widgets)		• Focus is on customized widget to meet individualized usages
	Provider Summary	Revenue = $60M Value Player–not the lowest cost, but solid products	Revenue = $39M leaving market– simply not enough profitability	Revenue = $18M Luxury Player–only provides customized widgets	Revenue = $42M Cost play–intends to expand into sport and build new markets	

BigGuyCo's exit. Over the following weeks, the strategic planning team assessed options until it eventually whittled down the list to two opportunities:

- *Opportunity 1: Expand distribution to former BigGuyCo retailers to capture market share in the general market.* BigGuyCo leveraged a handful of large retailers to distribute its products. WidgetCo's primary customer channel is to specialty shops via a distributor model. Surveyed customers were excited about the possibility of purchasing WidgetCo's products in more locations. The initial business case forecasted a sales pickup of roughly $25 million per year with a net present value of $45 million over the next five years.
- *Opportunity 2: Expand into entry-model sporting market to pick up BigGuyCo's market share.* BigGuyCo's offering in the sporting segment appeal to the beginning player. By introducing an entry-level sporting option, WidgetCo may be able to capture the market for this customer segment. This equates to a sales increase of roughly $20 million per year with a net present value of $38 million over the next five years.

From a competitive standpoint, Opportunity 1 would likely be met with a strong reaction from ForeignCo as it fought to retain its business in the mass market. With ForeignCo's superior resources, a frontal assault on ForeignCo seems ill-advised—especially when this segment of the market is declining. Additionally, opportunity 1 might change the WidgetCo's brand from a value player to a mass marketer and make it vulnerable in the higher-end sporting category.

Opportunity 2 expands on WidgetCo's strengths in the sporting segment and also helps it to get a foothold with new retailers.

ForeignCo may push to make a splash in the sporting market, but the spotty quality of its product and its lack of a reputation for sporting usages make such a move unlikely to succeed. During surveys, customers enthusiastically endorsed WidgetCo's possible expansion into the entry-level sports usage of widgets.

Based on this information, the strategic planning team adjusted its initial estimates to reflect the competitive risk of each approach. This resulted in a risk-adjusted net present value of $25 million for opportunity 1 and a risk adjusted net present value of $35 million for opportunity 2. Based on this analysis, the strategic planning group began building an initiative for opportunity 2. Because WidgetCo uses a process-based approach, the initiative-development process begins with the creation of a paragraph documenting the intent of the initiative

Initiative: Expanding Sports Offering

Based on the exit of BigGuyCo from widgets, WidgetCo will expand its offering in the sporting segment. This will include the development and launch of products aimed at individuals who are new to a sport (i.e., youth models and entry-point models). The new products will maintain the quality and performance reputation of WidgetCo. Additionally, the new products will require WidgetCo to expand its customer channels to larger retailers, where the bulk of the customers shop for these offerings. The consumer website will also carry these products.

Based on this initial description, an initiative owner was assigned the task of scoping the initiative and bringing it to fruition. Given the importance of the initiative and the need for timeliness, an experienced leader was asked to be the initiative owner. The initiative owner's immediate task was to identify the process requirements of the solution. Thankfully, the strategic planning team had already gathered the customer requirements (i.e., preservation of widget performance, entry-point product, and distribution through large retailers), so the focus was primarily on the internal processes to build the new offering.

STEP 1: PRIORITIZATION CRITERIA

The initial assessment of the initiative was performed by the strategic planning team and provided the ammunition to secure the agreement of the leadership council that the initiative should move forward. The leadership council at WidgetCo prioritizes initiatives based solely on their net present value as long as the resources are available and dependencies are accommodated.

STEP 2: PROCESS REQUIREMENTS

The enterprise process blueprint for WidgetCo is depicted in Figure 10.2. Based on a solid understanding of the enterprise's process structure, the initiative owner lists the processes requiring adjustments to support the initiative. This list is then confirmed by the affected megaprocess owners (i.e., distributor services, marketing, manufacturing, supply chain, website, and supporting processes) for accuracy and completeness. The list of process requirements looks something like this:

Initiative: Expanding Sports Offering

Distributor services

- Leverage team to initiate contact with larger retailers and begin discussions on carrying WidgetCo's entry-level sports offerings.

FIGURE 10.2 Enterprise process blueprint, Widget Division of WidgetCo.

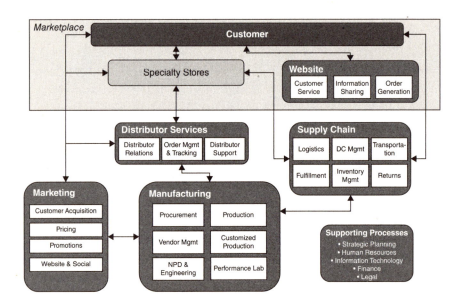

■ Examine feasibility of providing entry-level product through specialty stores as well.
■ Expand order-entry system to accommodate orders from retailers.

Marketing

■ Identify appropriate price points for new products.
■ Develop packaging for new products.
■ Plan promotional schedule to support launch of new products.

Manufacturing

■ Research and development creates specifications for new products.
■ Engineering expands manufacturing capabilities to build new products.

- Procurement sources materials required for new products.
- Performance laboratory tests new products to ascertain whether they meet expected quality standards.

Supply chain

- Develop new shipping routes to distribute products to new and existing customers.
- Estimate inventory required to keep up with demand.

Consumer website

- Work with marketing to update product offerings and product information on website.
- Support promotions.

Supporting processes

- Strategy estimates sales based on available market information.
- Information technology updates supporting systems to support the new product line.
- Human resources estimates new resources needs and prepares to hire and train new workers.
- Finance obtains the funds to support the new offerings and assists in developing the business case for the initiative.
- Legal develops contracts with new customers.

Process requirements will rarely cover all the effort required in an initiative, but the degree of detail allows the individuals responsible for execution to thoroughly understand its intent. Additionally, the process sponsors and owners understand what they need to do to make the initiative successful. Every well-designed initiative is sufficiently detailed to answer two questions:

- What are we going to do?
- How are we going to do it?

The process requirements more than adequately answer these questions. Additional details will be addressed once the initiative team is assembled. This is a good time for a checkpoint with the megaprocess sponsors (and process owners who own processes significantly affected by the initiative) to confirm the initial process requirements. At WidgetCo, the time designated to review process requirements occurs during a biweekly meeting of the leadership council. During this meeting, the "Expanded Sports Offering" initiative owner presents the process requirements and entertains questions. On occasion, the initiative owner might be asked to identify the key assumptions on which the initiative is predicated.

For the "Expanded Sports Offering" initiative, the following assumptions were previously identified:

- BigGuyCo exits the market, as stated in its press release.
- WidgetCo's internal team validates that capabilities exist (or can be created) to build the new products.
- Retailers will carry WidgetCo's entry-level sports offerings.
- The current channel (specialty stores) will accept the new offerings and still sell the existing sports offering.
- ForeignCo will not aggressively expand into the sports segment of the market.
- The expanded offering will be profitable and will not cannibalize any existing product lines substantially.

The assumptions are not intended to identify every possible incident or reaction that might occur—but they identify the major risks and assumptions on which the initiative's creation is predicated. When an assumption turns out to be false, the strategic implications and the business case need to be reassessed to determine whether the initiative should continue to move forward.

STEP 3: DEVELOP STRAWMAN SOLUTION AND BUSINESS CASE

The next step for the "Expand Sports Offering" initiative is to fully build the business case for the initiative. The business case is completed primarily by the initiative owner in partnership with a financial planning team and the affected process owners. To hasten the exercise, the work is broken down into two chunks—the cost to complete the initiative and the ongoing incremental cost-benefit once the solution reaches a steady state.

To identify the cost of completing the work, the initiative owner returns to the process requirements and identifies the process sponsors/owners to be involved in estimating the costs to deliver the outlined solution. Through meetings with each of the megaprocess sponsors and process owners, the initiative owner (in tandem with the financial planning team) calculates what it will take to adjust each process to deliver the new product offerings. And because it is a convenient time to discuss the work effort, an initial timeline for each chunk of work is built, including the identification of dependencies that need to be fulfilled prior to launching the initiative. This meeting might also cover the ongoing costs/benefits after the solution is implemented. However, because these estimates are likely based on the volume of throughput over time, these numbers will need to be reevaluated after the sales forecasts are finalized.

Table 10.1 shows the output of this exercise. The amounts listed and the time duration by megaprocess are provided by the megaprocess owner/process owner. Remember that these are only predictions, and their sole purpose is for comparing and ranking initiatives. Although accuracy is important, what is more important is to be consistent in the approach used to conduct the assessments. The financial planning team owns the methodology to build

TABLE 10.1   Costs to Implement Solution for "Expand Sports Offering" Initiative

	Estimated Cost	Timeline
Distributor Services		4 months
Retailer Acquisition	$200,000	
Expansion Specialty Stores	$150,000	
Marketing		1 month
Price Points	$30,000	
Packaging	$100,000	
Promotional Schedule	$200,000	
Website	$200,000	
Manufacturing		3 months
New Specs	$100,000	
Production Expansion	$100,000	
Procurement Sources	$50,000	
Performance Labs	$200,000	
Supply Chain		2 months
Distribution Network Assessment	$100,000	
Inventory Analysis	$500,000	
Supporting Processes		2 months
IT	$200,000	
HR	$40,000	
Finance	$2,000	
Project Team	$108,000	4 months
Total	$2,280,000	4 months

business cases and has the responsibility to maintain integrity in the overall process.

The "Initiative team" line item provided in Table 10.1 is an estimate based on the initiative owner's understanding of resources required to complete the initiative. For this initiative, a good portion of the work will be managed by process owners who use existing resources and processes. Still, there will be work to coordinate and manage the work efforts. The initiative-team estimate includes three full-time workers at a rate of $50 per hour for 18 weeks.

Knowing the cost to complete the initiative, the initiative owner now turns to the ongoing costs and benefits after WidgetCo expands its sports offering. The pro-forma financials depend on the sales volume of the new offerings. Creating these sales and cost estimates requires the participation of the strategic planning team (to estimate sales potential based on the available data and the predicted competitor responses) and the financial planning team. To allow for an apples-to-apples comparison among initiatives, the financial planning team selects the years to include in the business case and the discount rate. WidgetCo uses a five-year pro forma and a discount rate of 10 percent (Table 10.2).

For the purposes of simplicity, we will assume that the year 1 figure is for eight months after the initial four months to complete the initiative. The cost to complete the initiative ($2,280,000) then can be subtracted from the net present value (NPV) for the ongoing project to give us a total NPV of $27,307,081.

With the business case complete, the focus shifts to the initiative's dependencies and resource requirements. Because this is a new product offering, there are no immediate dependencies for the initiative. There may well be collaboration opportunities—especially with regard to further development of the relationships with retailers—but these will surface when we examine the overall innovation portfolio.

The next investigation is the limiting resources—that is, resources that are not readily available for use on the initiative. Limited resources differ across enterprises, but they always include the investment in dollars and resources to execute the initiative. We already know that the initiative requires three resources and roughly $2,280,000 over a four-month period. The initiative also requires the creation of new working relationships between different processes and stakeholders and the development of a new product offering. Anything identified as "new" provides a clue as to limiting resources. At first glance, it is impossible to tell if something is a limiting resource. What might be

TABLE 10.2 Ongoing Cost-Benefit for "Expanded Sports Offering" Initiative

	Year 1	Year 2	Year 3	Year 4	Year 5
Sales	$16,000,000	$19,000,000	$22,000,000	$26,000,000	$30,000,000
Cost of Good Sold	$12,000,000	$12,350,000	$13,860,000	$15,600,000	$16,500,000
Gross Margin	$4,000,000	$6,650,000	$8,140,000	$10,400,000	$13,500,000
Selling, General, Administrative Costs	$250,000	$305,000	$308,000	$320,000	$350,000
Profit	**$3,750,000**	**$6,345,000**	**$7,832,000**	**$10,080,000**	**$13,150,000**
NPV (at 10% discount rate)	*$29,587,081*				

limiting to one company (e.g., time from a specialized resource) may be widely available in another. Driving out the limiting resources requires further conversations with megaprocess sponsors and process owners.

In WidgetCo, a limiting resource is engineers to develop and build the production capabilities to manufacture the new widgets. The engineering department is constantly looking for new ways to improve the performance of its widgets because this is WidgetCo's strategic advantage. Depending on the allocation of the engineering staff to other initiatives, it may or may not be a constraint. In this instance, the engineers believe that they can fit the extra workload into their schedule without any issues. Besides the engineering team, the "Expand Sports Offering" initiative leverages existing processes and structures—although adjustments may be necessary as the sales volume increases.

With a complete business case and adequate knowledge of dependencies and limited resources, the "Expand Sports Offering" initiative can be inserted into the portfolio of initiatives and prioritized into the deployment schedule.

STEP 4: PRIORITIZE THE INITIATIVES

Adding the "Expand Sports Offering" initiative to the portfolio results in an updated table, as shown in Table 10.3. Currently, WidgetCo has two initiatives in flight—"Customer Behavioral Research Study" and "Marketing Expansion." These initiatives are being executed concurrently to take advantage of collaboration opportunities. Another initiative, "Widget Production Enhancements," is slated to start shortly. If we prioritize based solely on NPV, the "Expanded Sports Offering" initiative should be elevated to become the next initiative launched.

TABLE 10.3   WidgetCo Initiative Portfolio

Initiative	Risk Adjusted Contribution (000's)	NPV(000's)	Risk Factor	Resource Requirements		
				Budget $ in 000's	Team	Special Skills
Customer Behaviorial Research Study	$10,800	$12,000	10%	$1,500	Process Owner	NA
Marketing Expansion	$119	$125	5%	$750	Task Force	NA
Widget Production Enhancements	$5,220	$5,800	10%	$531	Task Force	Lean Skillset
Working Capital Assessment	$45	$50	10%	$150	Process Owner	Financial Analysis
Expand Sports Offering	$25,942	$27,307	5%	$2,280	Task Force	Engineering

Dependencies	Collaboration	Duration (months)	Major Process(es) Impacted	Delivery Status	Projected End Date
None	Marketing Expansion	2	Customer Acquisition New Product Development Marketing Customer Service	Green	10/25/2012
None	Customer Behavioral Research Study	6	Marketing Customer Acquisition Customer Service	Green	3/25/2013
None	None	6	Manufacturing (Performance Lab)	TBD	TBD
None	None	2	Finance	TBD	TBD
None	Marketing Expansion (?)	4	Marketing Manufacturing Supply Chain Distributor Services	TBD	TBD

In this instance, the leadership council needs to make a decision on when to launch the "Expand Sports Offering" initiative. In most instances, time is of the essence, and initiatives in process should continue without interruption as long as their resource needs do not preclude the launch of a vastly superior initiative. Time to market is critical for the "Expand Sports Offering" initiative, and there are no known resource constraints. It appears to be possible to continue with the ongoing initiatives and launch the "Expanded Sports Offering" initiative. Again, these are judgment calls based on the available information.

The next question to be answered is whether the "Widget Production Enhancements" initiative should be launched as well. Because there is significant overlap between this initiative and the "Expand Sports Offering" initiative, it becomes a question of the availability of knowledgeable resources. With two major initiatives focused on the same area, and with sizable potential benefits, deference should be given to the strategic initiative—the one most focused on building market share. Based on this rationale, the "Expand Sports Offering" initiative is slated as the next initiative to be launched. This approach allows the strategic adjustments to be completed before the process is made more efficient—thereby eliminating the risk that an outdated process (from a strategic standpoint) is improved only to have the efficiency efforts wiped out by the eventual strategic adjustment.

A final question to be answered is whether the "Working Capital Assessment" initiative can be launched. Again, if the resources are available and there is no impact on other initiatives, it makes sense to launch the initiative—especially because its work will be completed in finance, an area not heavily affected by other initiatives. With the prioritization complete, the next focus is on the available resources and the launch order of the initiatives.

STEP 5: SCHEDULE AND ALLOCATE RESOURCES

The last step in managing the portfolio of initiatives is to adjust the initiative launch order based on the availability of resources. Table 10.4 identifies the cash requirements for each initiative and the overall cash position of the enterprise with regard to the innovation portfolio. This same type of chart can be used to manage the availability of any limited resource. In this example, we would create a chart for the availability of engineering resources if there were a need for them outside the "Expand Sports Offering" initiative.

Based on Table 10.4, the "Expand Sports Offering" initiative cannot be launched because of the enterprise's limited cash resources. It will take one month to make a draw on the existing bank line. For this reason, the launch date for the "Expand Sports Offering" initiative will be pushed to November. The final activity in building the launch order for the initiatives is to account for any dependencies. If there are dependencies or collaboration opportunities, the launch order may be adjusted—thereby necessitating a change to the "Limited Resource Allocation Chart." Figure 10.3 provides a graphic view of the initiative launch order.

With creation of the "Expand Sports Offering" initiative and its insertion into the order of execution of initiatives, the portfolio is

FIGURE 10.3 Initiative deployment schedule.

Initiatives	Oct	Nov	Dec	Jan	Feb	Mar	Apr	May	Jun	Jul	Aug	Sep
Customer Behavioral Research Study												
Marketing Expansion												
Expand Sports Offering												
Widget Production Enhancements												
Working Capital Assessment												

TABLE 10.4 Limited Resource Allocation across Initiatives

(000's)	Oct	Nov	Dec	Jan	Feb	Mar	Apr	May	Jun	Jul	Total
Customer Behaviorial Research Study	$1,500	$0	$0	$0	$0	$0	$0	$0	$0	$0	$1,500
Marketing Expansion	$250	$100	$100	$100	$100	$100	$0	$0	$0	$0	$750
Working Capital Assessment	$70	$80	$0	$0	$0	$0	$0	$0	$0	$0	$150
Expand Sports Offering	$0	$540	$770	$540	$430	$0	$0	$0	$0	$0	$2,280
Widget Production Enhancements	$0	$0	$0	$0	$75	$200	$75	$75	$50	$56	$531
Total Cash Outlays (Innovation)	$1,820	$720	$870	$640	$605	$300	$75	$75	$50	$56	
Beginning Cash Available	$2,000	$180	$460	$90	$450	$345	$545	$970	$895	$845	
Cash Inflow (from Finance)	$0	$1,000	$500	$1,000	$500	$500	$500	$0	$0	$0	
Ending Cash Available	$180	$460	$90	$450	$345	$545	$970	$895	$845	$789	

set for the moment. However, as new information becomes available (perhaps BigGuyCo opts not to exit the market), the effects on the launch order need to be reassessed and the appropriate changes made to maximize the value of the innovation plan. As mentioned previously, the value of the innovation plan is maximized when initiatives are executed based on their net contribution, assuming that all dependencies and resource requirements are accommodated. Accomplishing this aim requires a continual review and assessment of the portfolio.

11

The Future of Process-Focused Enterprises

At an elementary level, innovation efforts are aimed at refining the true purpose of work efforts and ensuring that the work is done in the most efficient manner possible. This dual focus is at the heart of a process-based approach. Is the process fulfilling its purpose as intended (strategic-planning facet), and is it doing so as efficiently as possible (operational-innovation facet)? In this way, waste of all types—unneeded activities, lost time, and misused investments—is identified and eliminated. In effect, the enterprise calibrates itself to its mission by eliminating the unnecessary. This recipe for innovation opens the doors to a realm of possibilities.

Using processes as an organizational structure stretches the concept beyond prior usage. Contemporary organizational structures and managerial practices are remnants of ideas born from military conflict and the industrial revolution. Bludgeoned into their current form over years of use and misuse, the greater business community accepts their flaws to such an extent that engaging in a constructive dialogue on their practicality and utility requires the iconoclast to be armed with a comprehensive alternative. This book presents a novel approach for clearly illustrating operational components, directing work efforts, allocating resources, and tying all these pieces together through the language of process. For the sake of review, the basics of this approach are straightforward and simple:

- Develop a customer-focused mentality throughout the enterprise. Leverage internal and external data sources to gain a rich understanding of the customer's current and future preferences and behaviors. Consolidate this information in a format conducive to its dissemination to individuals responsible for developing tomorrow's product and service offerings.

- Use processes as the organizational structure to clarify the work performed in an enterprise and the interdependencies between operational processes (i.e., a process system).

- Implement a governance organization mapped to the process structure to ensure ongoing management and improvement of all areas of the enterprise. Assign process owners to focus on improving the efficiency of their areas and implement strategic adjustments in alignment with greater enterprise objectives.

- Plan changes to the overall process structure through an ongoing management of a portfolio of improvement initiatives. Prioritize and execute initiatives in an order that maximizes the total benefit generated by the full collection of initiatives and ensures that resources are allocated appropriately to support this innovation plan.

Although innovation can occur without this approach, it traditionally relies on leadership. And good leaders come and go—making it extremely risky to bet on leaders alone to steer any enterprise to greener pastures. The real aim of the approach presented in this book is to embed the approach, structures, and tools supporting innovation into the very DNA of the enterprise—thereby creating a system in which innovation is not simply a goal but rather an ongoing cycle that encompasses all the activities from opportunity identification to the delivery of a customer-desired product or service. In comparison with today's reality, where every improvement endeavor starts from a blank sheet of paper, the process-based approach to innovation is a framework for managing the continual improvement of an enterprise. Through its use, an enterprise surfs over the unending waves of change.

Corporations are the most immediate prospects for transformation to a process-based approach. After all, they have been the focal point of most improvement and innovation efforts over the past few decades. Results-driven like no other arena, market competition forces their hand to react to new market realities or perish. As their failures hit the headlines, corporate leaders see more emphatically than others the real risks confronting their companies. In small and entrepreneurial ventures, the risk is even greater. Without the safety net of substantial resources and market momentum, a process-based approach is particularly beneficial for smaller firms attempting to outmaneuver larger competitors. By focusing on the consumer's processes—how they shop, use, and service the product or service—the smaller firm can use this perspective to guide their improvement efforts.

Throughout this book, examples were presented of how this model applies to for-profit companies. However, I have been careful to use the term *enterprise* because a process-based approach is not restricted in its applicability. Bureaucratic and inefficient organizational practices are rife throughout all forms of enterprises but are especially prevalent in nonprofits and governmental entities. Lacking the push from the market, they are ripe for a transformation—and the potential benefits in such fertile ground are immense.

The process-based approach works in any type of organizational structure, including governmental agencies, nonprofits, educational institutions, and any type of nongovernmental or similar organization. Simply stated, if there is a mission to be achieved in a changing world (and it is an ever-changing world), a process-based approach is an efficient and effective approach to build a platform for innovation.

For example, imagine that governmental agencies and institutions deployed a process-based approach. The clarity itself and the ability to link actions to results would fundamentally change most governmental bodies. While I am skeptical that most governmental agencies would consider adopting a process-based approach because of the visibility it would provide to wasteful and inefficient operations and the resulting potential to hold leaders accountable for their actions, the long-term health of many levels of our government will require fundamental change to improve their operational capabilities. The day is fast coming when the U.S. government is severely crippled by the excesses and inefficiencies of past and current administrations. An approach that drives immediate efficiencies across the board may well be a significant opportunity to return stability to the U.S. government.

In a like manner, a process-based approach supercharges the nonprofit arena. With limited resources, the ability to focus resources on the critical processes to generate benefits for stakeholders allows nonprofits to accomplish their missions with fewer resources and less waste.

Or think of the benefits for educators and their students. Through the documentation and sharing of methods, approaches, lectures, coursework, and other educational elements, educators have empirical data to guide their teaching to optimize the learning, use, and retention of their students. But the benefits and expanded capabilities in the different forms of enterprises are only a toehold to the potential of the process-based methodology.

A process-based approach not only affects enterprises, but it also has the potential to reach down to the individual employees working across industries and geographies. With the rise and fall of the sun, hundreds of millions of people put in a solid day (or night) of work. The work is personal to the individual—not only a source of income, but it defines them in terms of their contribution to society and heavily influences their perspective of their self-worth. Inside most enterprises, individuals struggle with conflicting goals, poorly defined work elements, and a cacophony of confusion around the enterprise's direction. Job satisfaction is below average to poor according to many surveys. Employees perform their work in a perfunctory manner because they lack trust in their leaders and are unable to pinpoint exactly what their role is and what they should be doing.

In today's environment, promotions are rewarded based on others' perceptions of an individual's capabilities relative to other candidates. Performance cannot be evaluated on an apples-to-apples basis, and there is no foundational construct to discern the difficulty of any specific role. Process-governance structures, and indeed the whole process-based approach, remedy these deficiencies. Process complexity and performance can be associated and evaluated. Whereas the availability of information to evaluate leaders traditionally hindered succession planning, a process-based approach offers a wealth of data linked to specific directives and achievements of individuals. It is no longer possible for the do-nothing individual to speak well but perform inadequately and still get the promotion. The structure is in place to evaluate individuals based on their contribution to the work efforts of the enterprise.

Moving beyond individual and performance-management benefits, a process-based approach's potential to simplify and clarify work efforts is a game changer. Over time, the prevailing structures and approaches meld into business practices. Depending on their success and how widespread their adoption, these practices give birth to "best" practices. For example, with the current increase in technology implementations, a number of best practices were recognized for their ability to substantial boost the execution of technology programs. The Agile methodology is exactly such a collection of best practices. However, many of these practices have been challenged by practitioners for their lack of specificity or their applicability to other situations. Process models born of a process-based approach overcome these limitations and provide a previously unknown clarity of purpose. The detail afforded by a process perspective provides a crystalline view of a solution and an enhanced ability to register its applicability to other environments and situations.

In this way, process structures in the future may trace their roots to a collection of best-practice processes from different environments. The power of these structures is the ability to build new solutions—much like using blocks to create something new. Process blocks can be swapped in, swapped out, or reordered to deliver new outputs. Such a practice not only allows managers to leverage the experiences of others, but it also essentially functions as a shared language for change activities. The conceptual is made tangible. In effect, the potential exists to make the best methods and processes transportable from one environment to another—and at a specificity to make them implementable and meaningful. This facilitates a crisp flow of ideas—as well as provides foundational constructs that assist individuals in acclimating to new workplaces. This only comes about when the language of work is communicated via process.

This brings us to perhaps the greatest opportunity for extrapolating the capabilities of a process-based approach—the ability to map and innovate a full industry. Industries in their largesse are rarely considered from the perspective of the value they provide to society. Rather, they are perceived as a collection of individual entities (often corporations), each with a distinct offering and a unique manner of conducting business. Inside the walls of companies, corporate leaders aggressively seek advantages over their competition, and this effort results in industry inefficiencies and substantial opportunity costs. Competition has largely forestalled any efforts at cooperation and collaboration. As a result, companies spend money on research and development that is duplicative and would be unnecessary if they shared knowledge and processes. And every company builds a corporate infrastructure to support its business—spending valuable resources and energy on systems and processes outside the core value chain. They invest huge sums of money to promote their products and be top of mind to consumers. All this seems right to business and government leaders as they support the system implicitly by their inaction to change it. For a moment, look across the wall and see what might be.

What if companies shared back-office processes such as accounts receivable and accounts payable—allowing customers and suppliers a single point of contact for transactions? Just imagine receiving one consolidated bill from all your healthcare providers instead of multiple bills from individual specialists and other medical providers? Or consider the implications if enterprises syndicated to address shared needs. For example, what if companies invested into a knowledge-sharing portal—allowing for the elimination of duplicative experiments (such as those required for Food and Drug Administration approval of a new drug)? What if companies and other agencies built syndicates through a collaborative research and development effort to focus their resources to tackle major societal problems? What if government and nongovernment organizations worked in tandem with companies to build standards instead of allowing costly competitive fights to determine industry standards (like Blu-ray versus HD DVD in the high-definition DVD market) and make our overall society more efficient? What if we viewed an industry as a collection of enterprises and began managing the outputs of the industry in terms of value generated, safety, flexibility, and all the other process characteristics. By so doing, we leverage the enterprises' collective resources to deliver greater value for society. This powerful idea alone could fundamentally alter the way our society operates and the quality of life of every individual.

The benefits of this new manner of organizing and managing innovation are exciting and provide a distinct alternative to how enterprises operate today. The road to a better future begins with ideas—ideas that challenge current conventions and break rules. Processes have already demonstrated their ability to transform companies and industries. By providing an approach that leverages the power of process and uses it to drive innovation, the future is indeed promising.

Appendix
Process-Improvement Models

P rocess-improvement models are proven ways to organize work. The following models are employed in enterprises across the spectrum and are useful as fire starters—to help teams to consider alternative ways of designing and building processes.

1. Eliminate Functional Silos: Expand the Scope of Process across Functions

Description
Contemporary enterprises are plagued by organizational boundaries that inhibit the flow of both work and information from one part of the enterprise to another. This can be mitigated by expanding the boundaries of the process improvement effort to include a larger scope of work and eliminating the need to manage handoffs between different business units/functions. The benefit of this approach is a smoother flow of work across the end-to-end process. On occasion, legal or regulatory rules prevent the implementation of an end-to-end process.

Real-World Examples

- Management of automobile production lines from an end-to-end perspective
- Single owner of a customer relationship who owns all facets of the customer's interactions with the enterprise

2. STREAMLINE PROCESSES TO ELIMINATE OR REDUCE NON-VALUE-ADDING ACTIVITIES

Description

Evaluate the process steps in order, and eliminate any steps that do not add value or do not contribute to the value appreciated by the end consumer. Use this approach to ensure that the process is operating efficiently. Non-value-adding steps add cost and time. Even worse, they divert attention from activities that add value. As a precaution, always confirm the steps that are eliminated are not adding value with other stakeholders.

Real- World Example

- Reporting and confirmation steps are often needed at the inception of a process, but their value deteriorates over time because no one bothers to review the reports and make adjustments.

3. PERFORM WORK IN THE BEST PLACE AT THE BEST TIME BY THE BEST RESOURCE

Description

Match the performer with correct skill sets to the process at the correct time. Do not use an overly skilled resource to perform jobs that could be completed by another. In some instances, this means segmenting the work into what can be performed by one performer and pushing other parts of the process to a more skilled resource. Use this model when there is a variance in skill sets of performers, availability of resources, or availability of the facilities/equipment needed. One caution with this model is its inability to accommodate exceptions. The work performed must be fairly standard.

Real-World Examples

- Medical industry: Have nurses take vitals and basic information about a patient's condition prior to involving a doctor.
- Fast-food restaurants: Prepare food prior to its distribution to restaurants.
- Automobile servicing: Have apprentices perform simple jobs such as oil changes while seasoned mechanics address the more challenging jobs such as replacing an engine.

4. Single Performer

Description

Use a single performer to complete the work. The worker is responsible for all the steps in the process but may rely on business partners for inputs or added components to the end deliverable. The aim of this model is to reduce waste and inventory by allowing for a continual flow of work (i.e., consistent with Lean principles). This model is also effective when the process varies in inputs, outputs, or the steps in the process. An individual has the capacity to react faster than a team, and this eliminates the communication disconnects throughout the process. An alternative version of this model is to cross-train performers to enable them to play more than one role in a process, thereby increasing the flexibility to react to changes.

Real-World Examples

- Front desk service employee handles all forms of customer service but may hand off to others when necessary.
- Tool machinist completes process to create specialized components for an end product.

5. CASEWORKER/FACILITATOR PROCESSING

Description

This model uses a facilitator to manage the flow and execution of the process. Usually, this model includes the distribution of work to specialized performers and the return of the work when complete. The caseworker adds value by confirming that the work is completed correctly and that it flows to the next specialized performer for further processing. In some instances, the caseworker may execute components of the work. This model works well when the workflow requires adjustment for specific pieces of work (e.g., prioritization required on occasion to expedite certain customers' work) and when a number of specialized resources are required to complete the output. The keys to the success of this model are sufficient tracking of deliverables to allow the caseworker to monitor the flow of work and the availability of specialized resources to complete their component of the process.

Real-World Examples

- General contractor using subcontractors (e.g., architect, plumber, electrician, framer, etc.) to build a home.
- Private banker partnering with other banking specialists (e.g., investment adviser, mortgage banker, foreign currency buyer, etc.) to service a high-net-worth client.

6. ASSEMBLY LINE: SPECIALIZED PERFORMERS

Description

This model employs a number of workers, each of whom completes a component of work to build the end deliverable. A key contributor to the success of this model is the appropriate timing and execution of each task. This model works when the deliverables are standardized with minimal variance, when quality and consistency are prized in the deliverable, and when the performers are colocated.

Real-World Examples
- The manufacture of automobiles on an assembly line.
- Sandwich builders at fast-food restaurants, where each adds something to the sandwich.

7. CENTRALIZE

Description

Consolidate the performers of a process into a single functional area or geographic area to improve the performance. The aim of this model is to use economies of scale to drive down costs and increase the productivity of a group. Additional opportunities exist to drive down costs by eliminating overhead costs and driving down procurement costs. Use this model when there is limited to no variation in the process (e.g., standardization is possible) or there are limited resources (e.g., space, skill sets, machinery, technology and upgrade costs, etc.). When examining this option, always conduct a complete cost-benefit analysis of the opportunity.

Real-World Example

- Shared services (i.e., centers of excellence) to enable faster and more efficient processing of accounts receivable, accounts payable, sourcing, credit, and other functions.

8. Decentralize

Description

Use this model instead of centralizing processes when greater customization of the outputs is required, and there is a need for faster response times. In all instances, confirm the value of having rapid response or onsite solution delivery. In some instances, these are perceived as value enhancers but not valued by the customer. This model does increase the cost of the process because it requires the maintenance of additional locations in the field. Always confirm that the additional value delivered by field associates exceeds the additional cost by conducting a complete cost-benefit analysis including all additional costs.

Real-World Examples

- Staffing of loss-prevention specialists in the actual stores for large retailers.
- Onsite human resources assistance in manufacturing facilities.

9. Create Hybrid Centralized and Decentralized Organizations

Description

Fully centralized or decentralized organizations may not be the correct approach to serve the customer. There are instances where a hybrid solution may be best. The decision on what to centralize or decentralize often comes down to the specific situation. For example, a sales force may be allowed to set a price on certain bundles of goods but may require the input of a centralized pricing group for a more complex or specialized bundle. Use this approach when there are a set of consistent applications and a set of unknowns. The unknowns are best handled by a centralized group with access to greater information (i.e., across a larger population).

Real-World Example
- Servicing a banking customer—branch banking versus a specialized team.

10. Expand Product or Service to Create More Value for Customers

Description
In a highly competitive market, one competitive tactic is to expand the product or service to provide greater value for the consumer. This allows an enterprise to differentiate its offerings from those of the competition. When using this model, ask what customers value and what would make their experience more pleasant. Use this model when there are limited differentiators between the competitor's product and your own (i.e., product/service is a commodity). When using this model, make sure that the additional costs to deliver this tactic are covered by add-on revenue or new revenue streams.

Real- World Examples:
- Ability to book a hotel and transportation services when buying an airline ticket.
- Availability of delivery and installation services when purchasing a product.

11. Evaluate Whether Process Can Be Done More Efficiently in Another Way—Investigate Other Ways to Obtain Same Benefit or Function

Description

There are always alternative ways to complete the work performed by any process. By brainstorming new ways to provide the outputs, new processes can be created. Knowing what the customer wants and values, can you reengineer the processes to make the customer experience more valuable. Although this model is always appropriate, it is time-consuming and costly to evaluate every process using this technique. For this reason, use this tactic only on the most salient and valuable processes. On a smaller scale, process owners should continually reevaluate the processes in their area of responsibility with the goal of identifying more efficient ways to complete the same work.

Real-World Examples

- Apple reinvented the market for portable music through creation of the iPod and an online music store.

12. Limit or Expand Consumer Options to Ensure that the Value of the Offering Exceeds Its Cost

Description

This model segments customers to provide customized solutions. In some instances, the offerings are expanded to broaden the market appeal, and in other instances, the offerings are limited to focus on specific customer segments. Use this model when portions of the consumer market operate or purchase the offerings in a unique manner. In this way, pricing can be more closely correlated with the cost to service each segment of the consumer base.

Real-World Examples

- Limiting the available installation offerings of a contractor to focus on the most profitable installations (e.g., garage door openers and screen doors) while eliminating the more challenging and less profitable offerings (e.g., replacement windows).
- Providing additional options to ensure younger, more risky drivers at a higher cost and providing a standard rate for drivers over the age of 21.

13. Expand Process into Customer's Processes for Greater Efficiency

Description

In order to forgo the customer-queue bottleneck, certain steps of a process may be pushed to the customer for execution. This allows customers to perform these steps at their convenience and to perform them prior to the time they need to be completed—providing significant flexibility for customers. Use this model when a standard set of information is needed or activities are consistently executed and can be performed by the customer. If the cycle time to have the customer complete this step is less than or equal to the current turnaround time, having the customer complete these steps without assistance is an option. This process model commonly relies on technology to accommodate the interaction with the customer—bringing with it potential customer-service issues (i.e., lack of customer connection), lack of connectivity between initial contact point and the remainder of the process (i.e., having to provide information multiple times)—and it eliminates a customer contact point and the opportunity to understand and react to the customer's other needs.

Real-World Examples

- Self checkout or check in at hotels.
- Self-service customer-service call centers.
- Front-end Internet interfaces to collect information for a transaction (e.g., mortgage or opening a brokerage account).

14. Delay Process Decisions by Making Predictions of Customer's Need

Description

By making predictions about customer behavior, prework can be done on elements that are common to a product to allow for less overall throughput time, permitting the remaining decisions to be made at a later date. Use this model when there is a long cycle time for a product/service and significant variability in customer purchases.

Real-World Examples

- Fashion purchases: Ordering a number of products, but delaying the decision on the color and other select details.
- Fast food: Getting burgers on the grill but not placing the toppings (i.e., cheese, ketchup, etc.) until after the customer order is placed.

15. GO MANUAL

Description

Eliminate the technology supporting the solution to use a tried and true, more efficient manual process. Use this model when the technology is unstable or accommodates the process in an inefficient manner (i.e., the process inputs, outputs, or actual process has a high number of variations). The downside of the manual approach is losing access to the data an automated solution provides. The benefit of this approach comes when the costs associated with technology are removed from the equation. This approach is counter to the prevailing business mindset today, but its potential is significant.

Real-World Example

■ Manual workarounds when technology systems are unavailable or do not work as intended.

16. USE PARALLEL PROCESSES TO MINIMIZE DELAYS

Description

Often a process includes steps that could be executed concurrently to eliminate the wasted time (and money) in waiting for pieces of the output to be prepared. This model is applicable where there is no direct dependency between steps. When using this model, be sure to identify all the outputs of the activities that are to be executed concurrently and ensure that there are no dependencies between the concurrently operating activities.

Real-World Example

■ Processing a mortgage loan application and sending off information to multiple specialists to complete their part of the process.

17. AUTOMATE THE PROCESS

Description

The goal of automation is to reduce complexity by automating the performance of a process. This model facilitates consistent execution of the process and an expanded capture of data. When investigating this model, consider the following questions:

■ Does better information lead to a better result?

■ Is the collected information used, or can the mechanism to capture the information be eliminated (and the associated costs)?

■ If the information obtained was captured earlier and at a higher quality, would this be useful?

■ Is the information disseminated through the process to all potential destinations (including individuals outside the process and other business partners)?

■ Is the information available from other parties who could provide it and enhance the process?

■ Is the information captured in multiple places, and can the cost of capture be eliminated if the data was shared?

■ Is the information aggregated centrally?

■ Is the information analyzed at periodic intervals (including feedback loops from customers and frontline associates)?

When evaluating this as an option, always incorporate all costs (especially technology support costs, cost of downtime, upgrades, time to evaluate new data available, etc.) into the business case and account for the fact that automated processes may lose flexibility because any change may require software adjustments.

Real-World Example

- All enterprise packages, including enterprise resource planning, marketing resource management, and sales management programs.

Notes

Chapter 1

1. Peter F. Drucker, "The Discipline of Innovation." *Harvard Business Review*, May–June 1985.
2. "Reengineering the Corporation" by Michael Hammer and James Champy, published by HarperBusiness, New York, NY 1993.
3. Adam Smith, "The Wealth of Nations" published by W. Strahan and T. Cadell, London, England 1776.

Chapter 2

1. "The Boston Consulting Group on Strategy," published by John Wiley & Sons, Inc. Hoboken, New Jersey 2006. Original article entitled "Strategic and Natural Competition" written in 1980.

Chapter 3

1. Michael Hammer, "The Agenda" published by Crown Business in 2001 New York, New York.
2. Peter Keen, "The Process Edge" published by Harvard Business School Press in 1997 in the United States.

Chapter 8

1. Roy H. Williams, best-selling author and Marketing Consultant.

Index

About the Author

David Hamme is the managing director of Ephesus Consulting, a boutique-consulting firm based out of Davidson, North Carolina that focuses on driving game-changing initiatives for its clients. Prior to founding Ephesus, David worked as a management consultant for Ernst & Young and The North Highland Company and has served as an executive in Lowe's Home Improvement's $3B Installation Business Unit. As a leader at Lowe's, David oversaw the strategic planning, marketing, product management, pricing, new product development, and sales functions.

As a consultant, David's experience spans Strategic and Business Planning, Process Analysis, Design, and Management, Enterprise Cost Reduction, Performance Management, Program and Project Management, and Executive Coaching. Over an 18-year career, David has completed projects for over 30 clients including such recognizable names such as GE Capital, Kellogg's, Bank of America, Wells Fargo, Family Dollar, Delhaize USA, Fifth Third Bank, Lowe's Home Improvement, Time Warner, and Duke Energy.

David earned his BS in Industrial Management and Electrical Engineering from Purdue University, and an MBA in Finance from Indiana University's Kelley School of Business.